Systems Ecology

Systems Ecology

An Introduction to Ecological Modelling

R.L. Kitching

University of Queensland Press
St Lucia • London • New York

Typeset by University of Queensland Press
Printed and bound by The Dominion Press–Hedges & Bell, Melbourne

Distributed in the United Kingdom, Europe, the Middle East, Africa, and the Caribbean by Prentice-Hall International, International Book Distributors Ltd, 66 Wood Lane End, Hemel Hempstead, Herts., England.

National Library of Australia
Cataloguing-in-Publication data

Kitching, R.L. (Roger L.), 1945- .
Systems ecology.

Bibliography.
Includes index.
ISBN 0 7022 1813 8.

1. Ecology – Simulation methods.
2. Ecology – Mathematical models.
3. Ecology – Data processing.
I. Title.

574.5'0724

Library of Congress Cataloging in Publication Data

Kitching, R.L. (Roger Laurence), 1945- .
Systems ecology.

Bibliography: p.
Includes index.
1. Ecology – Data processing. 2. Ecology – Mathematical models. I. Title.
QH541.15.E45K57 1983 574.5'0724 82-20032
ISBN 0-7022-1813-8

Designed by Paul Rendle

To Mick Southern, Buzz Holling and Peter Geier
Teachers, Mentors and Friends

Contents

Tables

Figures

Preface

The extensive use of simulation models in ecology began in the early sixties and has gone from strength to strength since that time. Now, in the eighties, it must be considered a standard tool in many areas of the subject with a well defined role in both theory building and management applications. In many ways this development came too late for inclusion in the wave of ecological textbooks which appeared in the seventies. Admittedly, books of systems techniques have appeared such as that of Jeffers (1978) but these have focused more on the presentation of a set of techniques by mathematicians to biologists. Books about ecological modelling by biologists for biologists are few indeed. Gilbert, Gutierrez, Fraser and Jones (1976) presented their idiosyncratic but powerful approach to the study of insect populations in *Ecological Relationships* and, most recently, Berryman (1981) has provided us with a text on populations which takes a systems approach.

The present work is offered as complementary to these last two. It is an introduction to modelling at the behavioural, population, community and ecosystem levels and assumes no previous specialized knowledge of modelling on the part of the reader. It is for this last reason that introductory accounts of general systems philosophy and methodology, some mathematics and computing are included. For others, these introductory accounts can be skipped and the detailed examples presented as chapters 6, 7 and 8 (the major part of the book) approached directly. In choosing examples for inclusion, I have looked for cases at particular levels of abstraction or which used particular methods of modelling, which were also clearly presented in accessible publications. I apologize to the many whose excellent work was omitted, although not necessarily overlooked, and to those whose work I have described in detail, I regret any errors of misinterpretation that I may have made. Almost all of my accounts of particular models have been collated from published works, and interpolation or assumptions about implicit decisions made by the modellers have had to be made on occasion.

To acknowledge comprehensively all those who have helped me directly or indirectly throughout the long gestation period of this book is an impossible task. Some, however, must be mentioned because of the extent or very special nature of their contributions. Charles Elton and Mick Southern directed me to an ecological viewpoint which later metamorphosed directly into a systems one. Buzz Holling and Carl Walters, in particular, at the University of British Columbia, confirmed me in this transition and enabled me to become both systems ecologist and field naturalist. Peter Geier, Dick Hughes and Rhondda Jones, erstwhile colleagues at the CSIRO Division of Entomology, provided much of the groundwork that led to this work as well as specific information which has been incorporated in many small and not-so-small ways in the text. Myron Zalucki and Don Abel, at Griffith University, have spent many hours discussing issues, broad and narrow, and some of their viewpoints are inextricably bound up with those of my own expressed in this work. Bill Hogarth, Roger Braddock and Don Abel gave me specific assistance on matters of mathematical and computing detail; Myron Zalucki read and criticized an earlier version of the text; and Rhondda Row and Rosemary Lott acted as efficient research assistants when required. Judith Davies typed and typed and typed, version after version, and the final appearance of the work owes much to her uncomplaining efficiency. Merril Yule and Rosanne Fitzgibbons of the University of Queensland Press gave help and encouragement when it was most needed and saw the work through to completion.

The following individuals and organizations kindly gave permission to redraw or reproduce figures from works of which they are copyright holders: Academic Press (London) Inc. (figs. 31, 32, 33), Annual Reviews Inc. (fig. 19), Blackwell Scientific Publications (figs. 30, 31, 35, 36, 37, 38, 39, 40, 41, 42), Elsevier Scientific Publishing Co. (figs. 33, 71, 72), the Entomological Society of America (fig. 45), the Entomological Society of Canada (figs. 25, 26, 27, 28, 49, 50, 51, 52, 53, 54), W.H. Freeman and Co. (fig. 42), Professor M.P. Hassell (fig. 47), Dr G. Innis (figs. 65, 66, 67, 68, 69, 70, 71), Professor H.T. Odum (figs. 10, 71, 72), Dr J. Roughgarden (fig. 46), Springer Verlag (NY) Inc. (figs. 55, 56, 57, 58, 59, 60, 61, 62, 65, 66, 67, 68, 69, 70, 71).

To all of these people and organizations, and others unmentioned, I give my sincere thanks, while admitting that any error of omission or commission remains mine alone.

1

Systems and Systems Analysis

The study of ecology has gone through a number of phases since its inception with the botanical studies of Tansley and Clements. During the 1920s and 1930s, one strand of activity concentrated on the systematic and detailed description of the organism/organism and organism/environment relationships observed in nature. This was the "scientific natural history" school led by workers such as Charles Elton and Thomas Park, who were part of the great expansion in biological activity which followed the publication, acceptance and development of Darwin's ideas of natural selection and evolution. These workers were part of that new breed of biologists who looked for mechanism and explanation at all levels of study. It is still vital that ecologists, whatever their background and persuasion, should be able to describe what they see in nature, and to place interpretations on their observations which will help others to fit those particular building blocks into the overall construction which is the science of ecology. At the same time as the scientific natural history school was getting under way, a mathematical strand to ecology, notably in population studies, was beginning. This strand was influenced perhaps more by Thomas Malthus than Charles Darwin, and was a development of Malthus's early ideas that populations can be described usefully with simple equations of change. Workers from two continents, led by people like Verhulst, Pearl, Volterra and Lotka, were demonstrating that the growth and fluctuations observed in natural and artificial populations could be mimicked (or "modelled") by mathematical expressions, relating the rate of change in abundance of the organisms to certain quantities characterizing the reproductive behaviour of the organism and the environment in which they live. At this time few workers attempted to draw the natural historical and mathematical strands of endeavour together, although a notable exception was A.J. Nicholson. Early in his career Nicholson collaborated with the mathematician V.A. Bailey to carry out elegant work building mathematical models of the interaction between predators and their prey, identifying many key properties of the systems of equations that they erected. It is to their credit that much recent modern population ecology has its basis in the expressions and notions that these workers laid down.

The forties and fifties saw a flowering of theory in ecology but little mathematical underpinning for this theory. The major preoccupation of the period for many ecologists was the idea of the natural regulation of biological populations. Their observation of the natural world led workers to note that numbers of organisms in natural situations generally fluctuated about some mean level, and that there appeared to be forces in nature which maintained the population at about this level. The explanation of the regulating mechanisms involved led to sometimes acrimonious debates, which in many ways detracted from the solid foundation both in data, terminology and theoretical development which the period saw. In retrospect, the problem seems to have been that different workers erected different theories as to how population regulation might occur, but then generalized their notions, claiming a universality for the mechanisms they described which other workers were able to fault on the grounds that certain examples did not fit the theory. Accounts of the principles and disputes involved can be found in Clark, Geier, Hughes and Morris (1967), Price (1975) and Ito (1980). Other developments were proceeding in parallel with the population regulation dispute. The ideas of ecosystem and biome were being developed as a basis for the study of ecology following the lead set by the publication of the first edition of E.P. Odum's "*Fundamentals of Ecology*" (1953). The mathematical ecologists were productive also during this period but again tended to work somewhat in isolation from biologically based ecologists. However, there was a recognition among these workers at the time that sound data bases and the ability to apply their results in the real world were vital to the continued well-being of their subject. Particularly noteworthy in this area were P.H. Leslie (1945, 1948) whose elegant matrix models of populations are only now being used to the full (see for example the work of Usher 1972, Williamson 1972, Enright and Ogden 1979), and J.G. Skellam (1951) who showed that a complex ecological process, dispersal, could be approximated by the mathematics of physical diffusion. As earlier, very few workers appeared to span the gap between mathematics and biology. Lamont C. Cole (1954), with his treatment of effects of biological characteristics on performance at the population level, and fisheries scientists such as Beverton and Holt (1956) did manage to achieve this sort of synthesis.

The 1960s saw the arrival of the computer in the domain of most ecologists, and opened new vistas for some. For the first time, workers saw the possibility of handling the complexity that they observed in the natural world in a quantitative fashion, using the almost unlimited computational and storage abilities of large digital computers. Coincidentally, the period also saw a great flowering of interest and support for ecology, inspired in part by the growing realization of the burgeoning environmental problems of population, pollution and resource depletion. In many ways this was a fortuitous coincidence, as the ability to recognize and quantify the problems concerned depended in large part upon an extensive computing facility. During the sixties, the by now well established theoretical and practical lines of inquiry continued, as did traditional "mathematical ecology". However, C.S. Holling and K.E.F. Watt were among the first to see the value of the available techniques of computing and systems analysis in modelling ecological phenomena at all levels of resolution. Watt's "*Systems Analysis in Ecology*" (1966) was the first book to attempt any sort of synthesis in the area. Although necessarily limited in scope, this work served as an introduction and stimulation to a great many ecologists who later turned their attention to the systems approach. Watt next wrote "*Ecology and Resource Management*" (1968), the first work to cover the techniques, philosophy and applications of ecological modelling. Taking an independent but converging line, some workers were developing techniques of what has become known as "compartment modelling" (see chapter 8), originally to examine the fates of radiotracers in ecosystems (see, for example, Olson 1960, and the review by O'Neill 1979). This approach provided the foundations on which was built much later work at the level of the ecosystem, notably under the aegis of the International Biological Program (IBP). The IBP began in the late 1960s and the results from this mammoth set of collaborative activities are being synthesized using systems modelling techniques (for example Innis 1978, Breymeyer and Van Dyne 1980, Reichle 1981, Bliss *et al.* 1981).

The seventies have been a period of great excitement in ecology and, for the first time, significant numbers of people with adequate backgrounds in biology, mathematics and computing have been carrying out research on the widest imaginable range of topics. The monumental four volume compilation of Patten (1971, 1972, 1975, 1976) exemplifies the many aspects of the developing area that has become known as systems ecology. As well as detailed accounts of population and ecosystem models such as those of Hubbell on woodlice (1971), Williams on the lake ecosystem (1971), Wiegert on algal flies (1975), and Lugo, Sell and Snedaker on a mangrove system (1976), the volumes contain expositions on the philosophical aspects of the approach (Kowal 1971, Patten, Bosserman, Finn and Cole 1976), and on mathematical and engineering techniques newly applied to ecological questions (Gallopin 1972, Odum 1972, Waide and Webster 1976, Mulholland and Sims 1976). Patten's volumes have been followed up in many ways, notably by the series of edited volumes on statistical ecology, the later volumes of which cover topics in ecosystems modelling (see, for example, Innis and O'Neill 1979, Matis *et al.* 1979). In addition the now-thriving journal *Ecological Modelling* first appeared in 1975. This brief and selective history of the subject would not be complete without reference to the important essays of Geier and Clark (1976) and Pielou (1981), which present biologists' perspectives on modelling in ecology, pointing out the strengths and weaknesses, the promises and limitations of the methodology. It is in the cool grey light of these authors' comments that the present work is offered.

This book, then, presents an ecologist's view of some of the major techniques and modelling approaches in the application of the study of ecological problems at a number of levels (chapters 6, 7 and 8). These are prefaced by introductory accounts of modelling techniques (chapters 2 and 3), computers and computing (chapter 4), and some key aspects of mathematics (chapter 5). These will provide the reader with sufficient information to handle the descriptions given in later chapters and will provide entries to more advanced literature. The concluding chapter gives a brief account of prospects and pitfalls in systems ecology (chapter 9).

It should be stated at the outset that a broad view is taken of the term "systems ecology", covering areas of population biology and behavioural ecology, as well as community and ecosystem study. This contrasts with the accounts of some authors who equate "ecological systems" with "ecosystems" and, accordingly, restrict their use of the term "systems ecology" to modelling approaches to ecosystems only (for example, Wiegert 1975, Innis 1979, Shugart and O'Neill 1979).

Some areas of importance to the ecologist interested in modelling are not covered in this book. Most important of these is a treatment of the statistical techniques which must be applied to data collected as input to a model, on the one hand, and to the predictions of a model and their comparison with verifying data, on the other. This omission is not considered disabling, however, as there are many excellent introductory works on the subject. Jeffers (1978) provides a particularly lucid overview of certain aspects of the area of particular relevance to the would-be modeller. This work also deliberately excludes the mathematics of dynamical and control systems. The interested reader is directed to the works of Rosen (1970) and Milsum (1966) for introductions to these areas of endeavour.

The view presented here is necessarily a personal one; other authors would undoubtedly have chosen a different set of topics or different slants on those that are included. However, I believe the material presented makes up a coherent whole which will provide the student of ecology with an introduction to and taste for a rapidly developing and powerful approach to his or her subject.

The remainder of this chapter explains the basic philosophy of systems ecology and its relationships with the natural world and the rest of ecological science.

1.1 What Is a System?

Like many words, "system" is elusive when it comes to exact definition. Webster's Dictionary includes the description, "an aggregation or assemblage of objects joined in regular interaction or interdependence", as part of a very long entry under the word. Other lexicographers provide variations on this theme, but common to all their ideas is the notion of a set of objects or **components** which **interact** together in space and/or time. An additional characteristic inherent in the term as we shall use it is that a particular group of components and their interrelationships has been **chosen** for some purpose — to answer a particular question, illustrate a theory or in an attempt to classify part or all of the natural world.

Being subjective entities, systems have two properties which make them particularly useful concepts in the study of ecology:

1. They may be **nested** — thus the individual is a part of the population of a particular species; that population, with others, is a part of a community of organisms; the communities may be organized into ecosystems and so on. At each level of resolution the organism or set of organisms

concerned, together with a greater or lesser part of their surroundings, may be regarded as a system. Indeed, one can regard the whole physical universe as forming a continuum of systems from the sub-atomic to the astronomical levels of resolution, and any of these levels can be studied using a common set of principles and techniques often referred to as **General Systems Theory** (von Bertalanffy 1964).

2. Systems at the same level of resolution may **overlap**; for example, the fox is an important predator upon both rabbits and beetles and, in the study of the populations of either sort of prey, the fox would be included as a component of the system of interest which is recognized around each. But would the rabbit be included as part of the system defined around the beetles or vice versa? Of course, the answer would depend on the exact purposes of the study, but would probably be "no" if the primary intent were understanding the population dynamics of the particular prey species. Foxes are a component of both systems

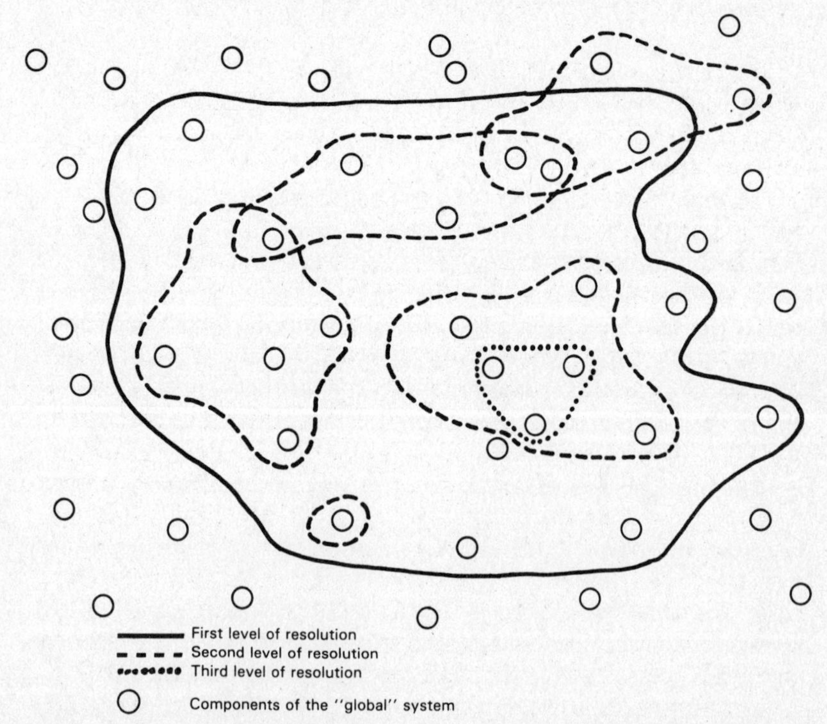

Fig. 1 Schematic representation of nesting and overlap in systems of components

which overlap at this point. This raises the further problem of defining the boundaries of the systems being studied, which will be considered in chapter 2. Figure 1 illustrates the notions of nesting and overlapping in systems.

1.2 What Is an Ecological System?

Perhaps the most general although not necessarily the most useful definition here is that ecological systems are those chosen for study by ecologists! Again, the subjective nature of systems definition is underlined in this tautology.

In practice, ecologists have tended to work at one or more of a few basic levels. The most ambitious have been those who have studied the **biosphere** as a whole (for example, Meadows *et al.* 1972, Forrester 1971, Mesarovic and Pestel 1975), attempting to characterize and make predictions about the whole of life on this planet, centering on man's present predicament. The principles of this approach are essentially the same as those used by ecologists studying somewhat smaller portions of the global system.

The **biome** and **ecosystem** have been the levels of resolution for the recent massive effort by ecologists within the International Biological Program, and a great deal of work synthesizing these studies is now appearing. Patten's Volume 3 (1975) includes summaries of several of these, specifically referring to the grassland (Innis), deciduous forest (O'Neill), desert (Goodall), tundra (Miller, Collier and Bunnell) and coniferous forest (Overton) biomes, together with a useful commentary by Watt. The model constructed of the grassland biome in North America has been reported in great depth by Innis (1978), and is considered in detail in chapter 8. Other workers have chosen the ecosystem as the level of resolution for their research also, notably the Odums in their characterizations of ecosystems such as Eniwetok Atoll (Odum and Odum 1955) and Silver Springs, Florida (Odum 1957). In addition, H.T. Odum and his students have examined several systems using an "energy circuit language" reminiscent of the circuit symbolism used by electrical engineers (see Odum 1971, 1972 and chapters 2 and 8 below). Many of the developments emerging from this school have been collected in a volume edited by Hall and Day (1977).

The **population** is probably regarded as the basic unit of study by a large proportion of animal ecologists, so this level of resolution has received much attention from the ecological modellers. Lotka (1925), Volterra (1926) and Gause (1934) were pioneers in the field, producing simple but extremely influential mathematical models of competition and predation at the population level. Nicholson and Bailey (1935) and, more recently, Hassell and Varley (1969), Hassell (1978), and May (1981*c*), among many others, have expanded and refined the earlier models, introducing varying degrees of additional complexity into the same basic type of model structure. An alternative development of especial importance in applied ecological studies has been the "life-system" approach of Clark *et al.* (1967), where the system of interest is defined as the population (or part of it) of the subject species, together with those elements of the environment which affect or are thought to affect the numbers of the organisms concerned. This way of looking at a species has been widely employed by entomologists, and the studies of aphids by Hughes and Gilbert (1968), Gilbert and Hughes (1971) and Gilbert and Gutierrez (1973) provide a very useful introduction to the power and usefulness of the methodology; a recent general account of the approach is given by Clark, Kitching and Geier (1979).

Lastly, the individual organism, and those items of the environment that impinge upon it, have been defined as a system for detailed study by a number of ecologists primarily interested in **processes** within the population. From a scientific point of view, the separation of this level of approach from that which looks at the population itself is perhaps less than wholly desirable. Workers who have studied processes by characterizing the behaviour of individuals are interested in the average behaviour of members of biological populations. Here ecology and ethology meet and no rigid distinction is possible or desirable between the two. The work of Holling and his co-workers (Holling 1959, 1965) on the process of predation may be taken as a model of the approach, although other more restricted studies of the processes of competition (Griffiths and Holling 1969) and movement (Kitching and Zalucki 1982) have been made in a similar fashion.

Examples of successful studies at the level of the population, individual and ecosystem are presented in chapters 6, 7 and 8. However, it cannot be overemphasized that these approaches are nested and hence not independent of each other. Each approach attempts to answer particular classes of questions, but the worker at one level must be aware of the developments made at the other levels. Indeed the strata identified are usefully regarded only as segments of the continuum of natural systems already alluded to, and ecologists can and do choose intermediate levels when the questions they wish to ask warrant this choice. Any choice within the continuum is necessarily an artificial one and it is most important that this artificiality be recognized and borne in mind when interpreting results.

1.3 What Is Systems Ecology?

Systems ecology can be defined as the approach to the study of the ecology of organisms using the techniques and philosophy of systems analysis: that is, the methods and tools developed, largely in engineering, for studying, characterizing and making predictions about complex entities, that is, systems. Beyond this, every worker in the field might give different details of what should be included under the term "systems ecology"; nevertheless, the term has come to have wide usage, and I will look more closely at how it is used or might be interpreted.

Ecologists have always been interested in complex systems, and one of the finest and most complete studies was directed by Charles Elton, and summarized in his seminal book, "*Pattern of Animal Communities*" (1966), in which he describes thirty years work on an area of deciduous woodland in Berkshire, Wytham Woods. Although most would not classify Elton as a systems ecologist, he did choose to look at a portion of the natural world, designated as the system of interest, recognizing its basic components and the interactions among these components and with the outside world, and then studying this subject set of the world in detail. This procedure is precisely that followed by a systems ecologist. The fact that Elton would not normally be included under the recent term, systems ecologist, is because it has come to be associated with a particular tool, the computer, and the technology which has built up around it. This distinction is unfortunate in many ways as it is the philosophy of the approach to the study of ecology that is important, and it would be fallacious to regard recent developments as basically new in this regard. However, Elton's synthesis was a 432 page book, his verbal model, whereas the modern worker is able to use the computer to convert such verbal or mental models into analytical or numerical ones, which can be studied and manipulated as analogues of the real system. This capacity has led to the formulation of a set of semi-formal techniques and rules, based on the capacity and mode of operation of the computer and the requirement that regular recourse must be made to the real system to verify and test the models generated.

The existence of a wide range of environmental problems and the increasing interaction between man and his environment has led to a recent emphasis on the need for development of management techniques as a most important part of applied ecology. The great value of simulation models of natural systems for assessing management alternatives has been realized, and this realization is evidenced by general works such as those of Russell (1975) and Welch and Croft (1979). Other authors have developed models for evaluating strategies of pest control (for example, Stark 1973, Blood *et al.* 1975), fishing activities (Paulik 1969, Larkin 1971, Walters 1969) and wildlife management (Walters and Bandy 1972), among a wide range of other applications. Any real distinction of course between "pure" and "applied" sides of any scientific endeavour is more imaginary than anything else, the notions representing entirely complementary facets of the same thing. Nevertheless, it is possible to pick out examples of work done in a strictly problem-solving milieu which draw upon the systems ecological approach. Indeed the approach is particularly well suited for the complex problems of environmental management, and may be expected to be one of the main developments within and around systems ecology in the future.

Computer technology has blossomed (some might say exploded) in the last decade or so, and there can be few areas of our lives in which the presence of the computer cannot be detected in some form or another, from the prediction of the results of elections, to lottery draws, to the processing of electricity bills. This, unfortunately, has led to the build-up of a mystique around the "magic machine", and this mystique is also present, albeit in a more subtle form, within the ecological community. A considerable amount of scepticism has been expressed by some workers with regard to the large and not so large numerical and analytical models built and used by some of their colleagues. It is one of the primary aims of this book to dispel the mystique and show systems ecology for what it really is: a useful and general way of looking at the natural world with a powerful and readily accessible technology to back it up.

2

Describing the System

In describing ecological systems it is most convenient to adopt standard terms and methodologies. Systems ecology has taken to itself a terminology and descriptive mode derived from usages in the engineering and mathematical sciences, and this chapter introduces two of these areas — systems description and systems diagrams.

2.1 Systems Description

In any study of an ecological system, an essential early procedure is to draw a diagram of the system of interest. Figures 2 and 3 are examples of such diagrams: figure 2 represents the processes of energy-partitioning in a predatory bug and figure 3 represents the population dynamics of the bug. Both diagrams are based on work carried out by G.C. Gallopin and myself on the North American species, *Podisus maculiventris*, but for our present purposes may be regarded as exemplars for any system that the ecologist may wish to study. For this reason, the diagrams have been kept as simple as possible: the energy-partitioning model represents the processes in the male bug, thus obviating the need for a reproductive component of any significance; at the population level, predators are assumed to have no significant effect on the numbers of their prey.

The diagrams indicate the systems' **boundaries** by a solid line. Within these boundaries, series of components are isolated which have been chosen to represent that portion of the world in which the systems analyst is interested — in these particular cases the food processing system and the population dynamics of a predator. If there are no connections across the systems' boundaries with the surrounding **systems environments**, the systems are described as *closed*. Ecological work, however, deals almost exclusively with *open* systems (the exceptions being in

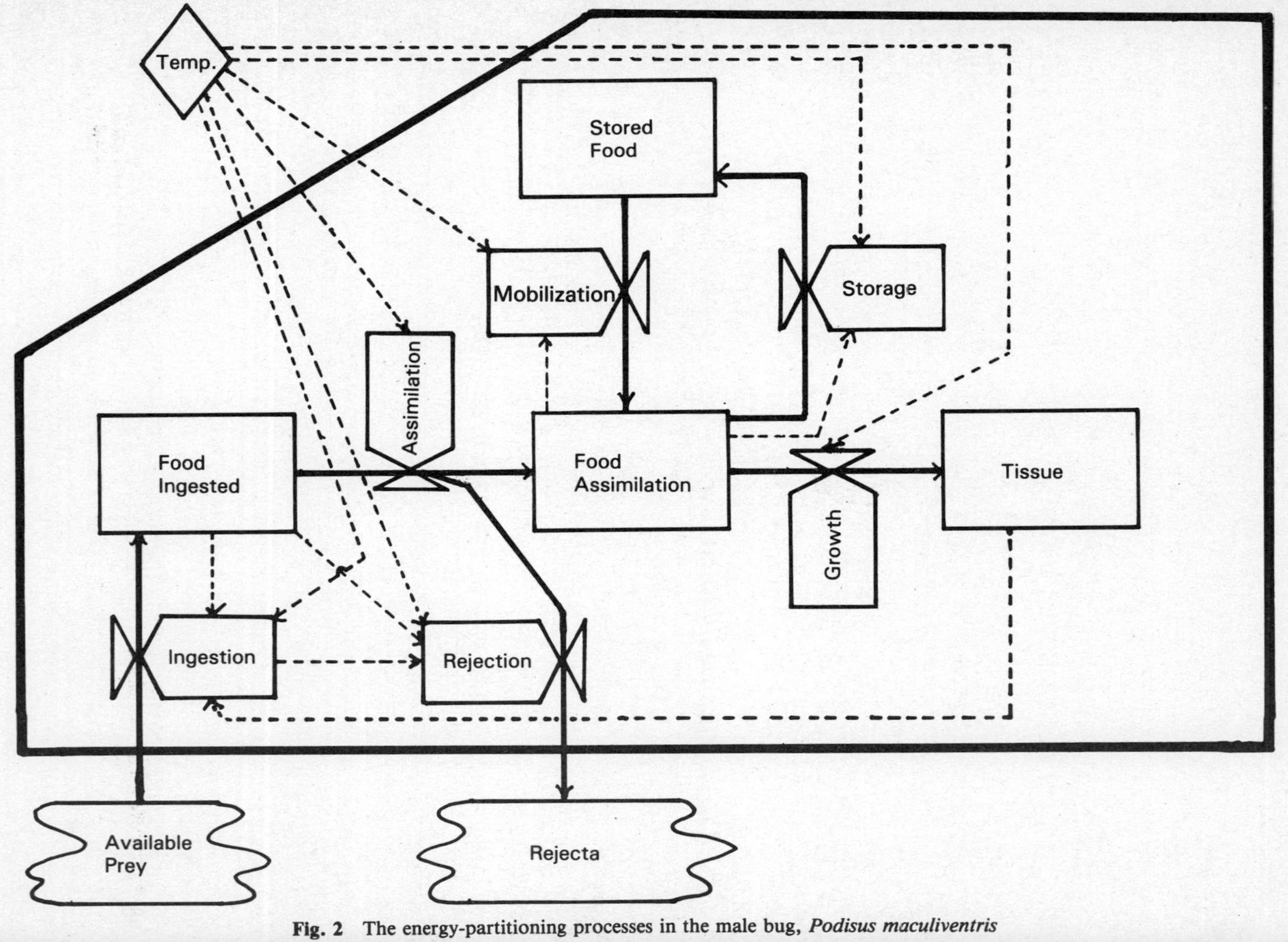

Fig. 2 The energy-partitioning processes in the male bug, *Podisus maculiventris*

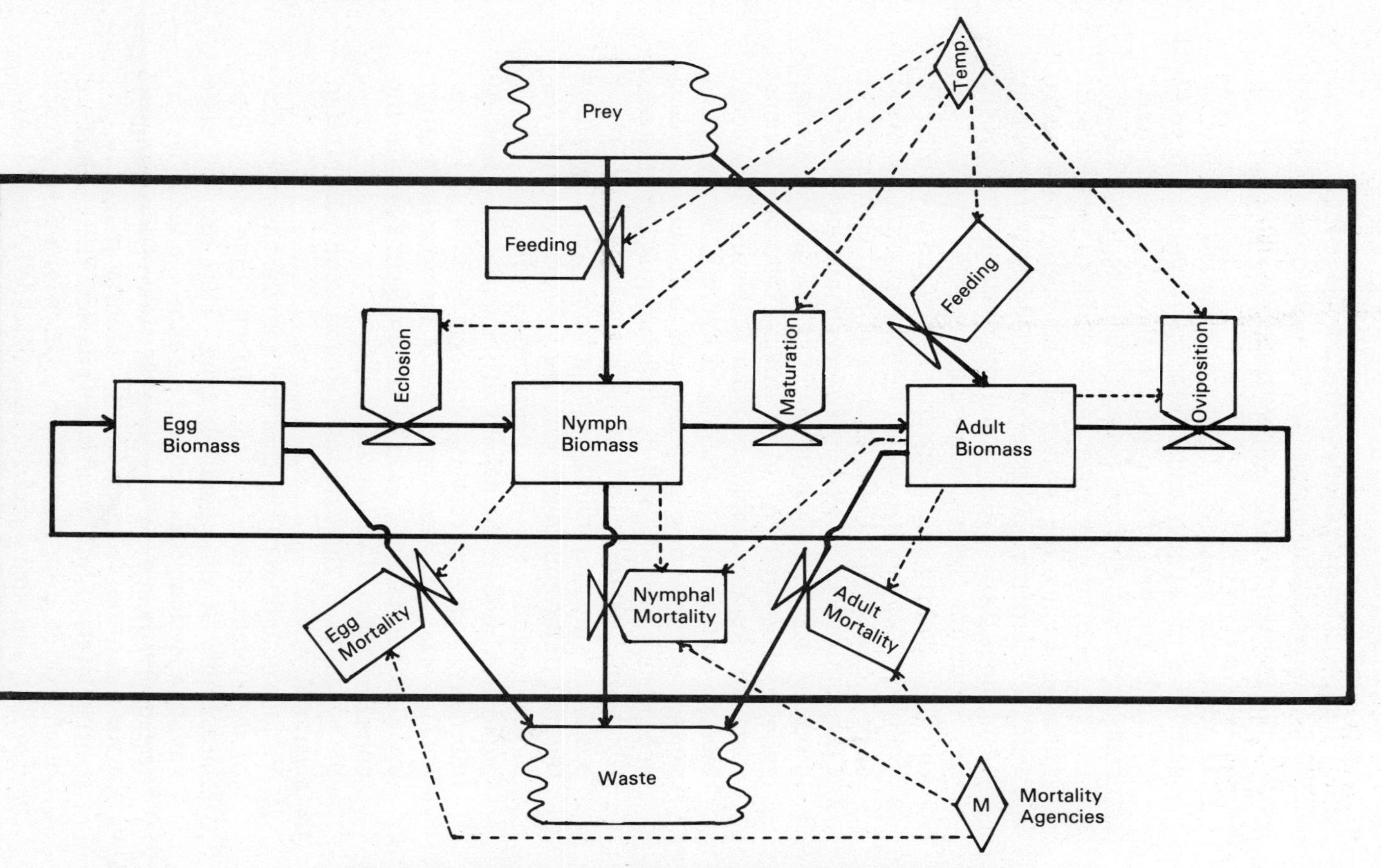

Fig. 3 The dynamics of populations of the bug, *Podisus maculiventris*

certain theoretical investigations where, for the purposes of the study only, an ecological system can be regarded as effectively closed). In open systems, two categories of phenomena outside the systems' boundaries are of special note:

1. There are the **sources** and **sinks** of materials which enter and leave the system and which are themselves beyond control by any part of the system. In the examples, material entering the system from outside is derived, in both cases, from the pool of prey organisms available to the predator. Material leaving the system in the energy-partitioning example enters the pool of rejecta made up of faeces and nitrogenous wastes from within the predators' bodies. In the population system, material leaves the system and enters the detritus sink if material is removed as a result of deaths other than those caused by predators and parasites, or is incorporated into the biomass of mortality agents.
2. Certain quantities outside the systems boudaries will affect processes within them profoundly. These are the **driving variables** (also referred to as driving functions, forcing functions, input variables or input functions). In a cold-blooded animal such as the bug in figure 3, the most important of these driving variables, at both the individual and population level, will be environmental temperature, which will "drive" all the physiological and behavioural processes within the animal without being itself affected by the state of the animal (see Kitching 1977 and chapter 6 for further details of this dependence). In the simple population model, the actions of the various agencies of mortality, such as predators, inclement weather and so on, are also regarded as driving variables having a direct effect upon the mortality of each of the different stages of the animal recognized within the model.

In the system itself, three classes of items can be recognized in the diagrams:

1. Various quantities of material make up the basic structure of the system; these will vary through time but will be measurable at any point in time. They are the **state variables** (system variables) and are shown in the figures as rectangular boxes. At any time the state of the complete system can be represented by numbers indicating the current **levels** of this set of state variables. These values are often written using the vector notation of linear algebra (see 5.2), and such a representation is called a **state vector**. In the examples, the system shown in figure 2 can be

represented by the levels of the four state variables: food ingested (I), material assimilated (A), stored food (S) and tissue (T). In the second example, the population model, the comparable variables are egg biomass (E), nymph biomass (N), and adult biomass (P). For each system, then, the set of state variables at any time, t, can be represented as vectors:

$$[I,A,S,T]_t \text{ or } [E,N,P]_t$$

In any system, each state variable is designated for the particular level of resolution being used in the study. For example, the box labelled "stored food" in the energy-partitioning model takes no account of the form in which the food may be stored nor of any internal rearrangements which may occur within the box. Similarly, in the population model the box labelled "nymph biomass" takes no account of the fact that there are five separate stages making up that period of the bug's life which is refered to as the nymphal stage. Each could be expanded into a complex system of interacting components in its own right, but at a *lower* level of systems resolution in each case. However, at the level of study selected, we are not interested in the detailed internal workings of the system of food storage in the one case or the nymphal biomass in the other, but only in the fact that material enters and leaves the box according to certain quantitative rules which may be determined experimentally. Thus, the component is often referred to as a **black box** (see Ashby 1956 for a more extensive account).

2. Figures 2 and 3 show arrows of two kinds connecting the state variables with each other and with the sources, sinks and driving variables outside the system boundary. These are the **inputs** and **outputs** of the components or, at another level, of the whole system. They represent **flows** among the parts described and may be of one of two types, material or **information**. Information is a difficult term to explain briefly, but can be regarded most simply as a signal from one component to another, telling the receiver something about the state of the transmitter without any material transfer being involved. The quantification and manipulation of such signals is the stock of information theory, and introductions to this topic are given by Shannon and Weaver (1949), Burch and Strater (1974) and, in an ecological context, Margalef (1957).
3. Material flows between components are regulated by **rate**

processes, represented in the figures by boxes "pinned" to the appropriate flow arrows. These processes will bear functional relationships to some of the state variables and some or all of the driving variables. The speed of action of each process will be modified by information flows from various other components. The rate of ingestion shown in figure 2, for example, is functionally related to the state variables, tissue and food ingested, and the driving variable, temperature. This is because the amount of food taken in, given a situation of unlimited availability of prey, will vary with the size of the animal and its recent feeding history, as well as with the ambient temperature. In the population model, the rate process labelled egg mortality may be taken as an example. In any time period, the number of deaths in the egg component of the system will depend upon the number of eggs, the number of nymphs and adults in the vicinity which may cannibalize the eggs, and the levels of the various external mortality agencies such as egg predators, parasites and violent weather which may wash eggs from their position.

The relationship shown in figure 2 between the amount ingested and the rate of ingestion is an example of a **feedback** loop within the system and, in general, such a loop can be represented as shown in figure 4. Feedback can be either **positive** or **negative.** Both forms occur in ecological systems although the negative variety is more commonly identified and discussed. Feedback is described as negative when it reduces the strength of a rate process. Returning to the processes of ingestion in figure 2, the rate of food intake will diminish as the crop of the insect becomes progressively fuller. A good example of positive feedback is found in figure 3, where the number of eggs laid will be positively affected by the number of adults in the population. In an unrestricted situation this will lead to the well-known phenomenon of logarithmic growth of the population in an effectively unlimited environment. To put it another way, the rate of increase in numbers is largely a function of the size of the population; as this increases, so does the rate of increase. In other words the process "snowballs" through time.

The concept of feedback leads us to the very important notion of **control.** Control implies that there is some normal or "preferred" value or level for a particular state variable, such as respiratory rate or body temperature in

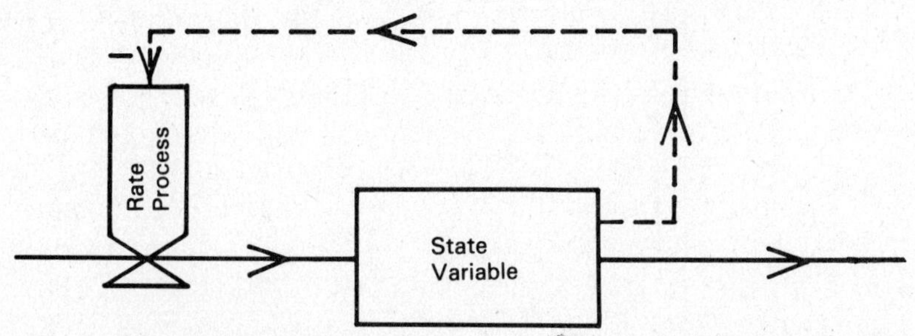

Fig. 4 A negative feedback loop in which the level of a state variable (SV) has a reducing effect on the rate process (RP)

warm blooded animals, net reproductive rate in a population system or species diversity in a biological community, and that the system acts so as to restore these variables to their normal values following any **pertubation.** Such control may be direct and **monotonic** as in figure 5, or compensatory or **oscillatory** as in figure 6. The former type is seen commonly in physiological processes such as some of those already mentioned, whereas oscillatory control is the sort of process observed when a population of animals in a stable environment is perturbed (as for example, in laboratory cultures of some insects). Control of course may not always strive to maintain a particular level in a state variable, but may simply restrict it between an upper or lower bounding value. In this situation regular, stable **limit cycles** may occur, in which the level varies in an oscillatory fashion with a constant period and amplitude.

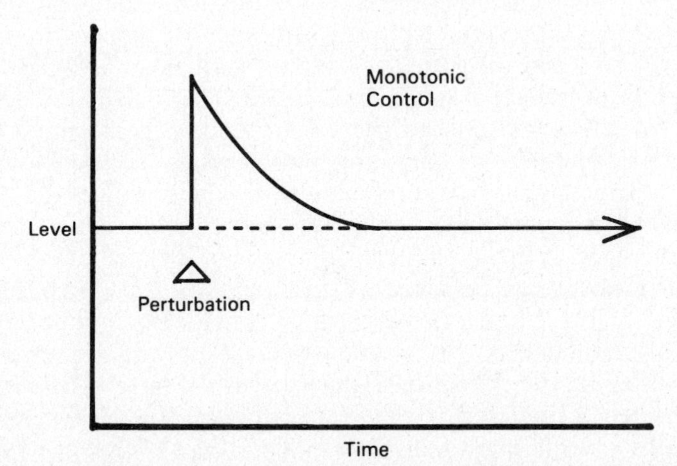

Fig. 5 Monotonic control of the level of a state variable following a pertubation

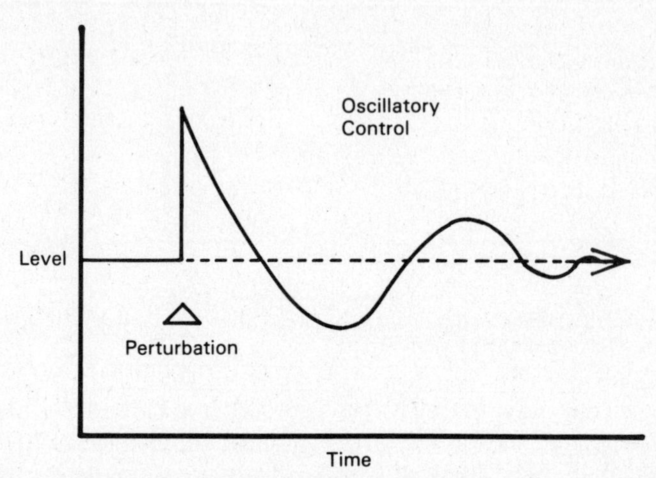

Fig. 6 Oscillatory control of the level of a state variable following a pertubation

The description of systems in terms of state variables, rate processes and flows is, essentially, a **static** characterization. However, the representation of the systems in terms of state vectors enables us to approach their **dynamics** through time. The vector $\mathbf{s}_t$, describes the state of the system at time t. At a later time, say $(t+\Delta t)$, the system will be described by the vector $\mathbf{s}_{t+\Delta t}$. The change from $\mathbf{s}_t$ to $\mathbf{s}_{t+\Delta t}$ is the basic system **transition** or **transformation** and can be represented most succinctly as a matrix multiplication:

$$\mathbf{s}_{t+\Delta t} = \mathbf{T}\mathbf{s}_t$$

Where **T** is the transition matrix describing mathematically how each element (state variable) in the state vector changes through time. The use and mechanics of matrix and vector representation will be described in more detail in 5.2. The changes inherent in the transition will depend upon the functions represented by the rate processes. In these functions certain constants will occur, characteristic of the system as a whole, and these are referred to as **parameters.**

With regard to the dynamical behaviour of a system, two further distinctions can be made which may allow the system as a whole to be described more accurately. These depend upon the kinds of mathematics used in the representation of the system. In speaking of the transitions which occur we have considered changes over a time step, $t \rightarrow t + \Delta t$, and this sort of approach means we have adopted a **discrete** model of the system in which we examine the behaviour at a series of points in time. This way of looking at the system is particularly well-suited to the use of the digital computer and to mathematical models couched in terms of difference equations. The alternative is to construct a **continuous** model of the system in which we describe changes using the infinitesimal calculus and the descriptive power of differential equations. This enables us to look at the way a system behaves through time rather than in a stepwise manner. This approach is most closely served using the analogue computer, although special approximation procedures allow the use of digital machines in this area also, with virtually no loss of precision or convenience (see 4.1).

In addition to their discrete or continuous nature, models of systems will be either **deterministic** or **stochastic** (probabilistic), depending upon the nature of the transition processes involved. If these always give the same answer, given the same starting state, then the model is said to be deterministic. If, however, the transition relationships are defined in terms of probabilities, then different answers may be obtained in several calculations of the transition, even with the same starting conditions. In this case the model is said to be stochastic and calculations are frequently made with the aid of a computer programme which generates "pseudo"-random numbers. An introduction to the basic ideas of probability is given in 5.4. The simple model of predation developed by Holling (1965) is a good example of a deterministic model, and that of dispersal developed by Kitching (1971*a*), of a stochastic model. These examples will be examined in more depth in chapter 7.

2.2 Systems Diagrams

In the preceding section, the terminology of systems ecology was introduced using diagrams (figures 2 and 3). The symbology used in these diagrams was a simplified version of that proposed by J.W. Forrester (1961). In this section, I intend to formalize and complete the introduction to the Forrester method of diagramming systems, and to introduce a second system, that of H.T. Odum (1972). These are but two of the great variety of ways in which such diagrams may be constructed, and have been selected simply because they enjoy considerable currency with ecological modellers. The major difference between the two in ecological terms is that the Forrester system allows us to represent flows of materials and information from one part of a

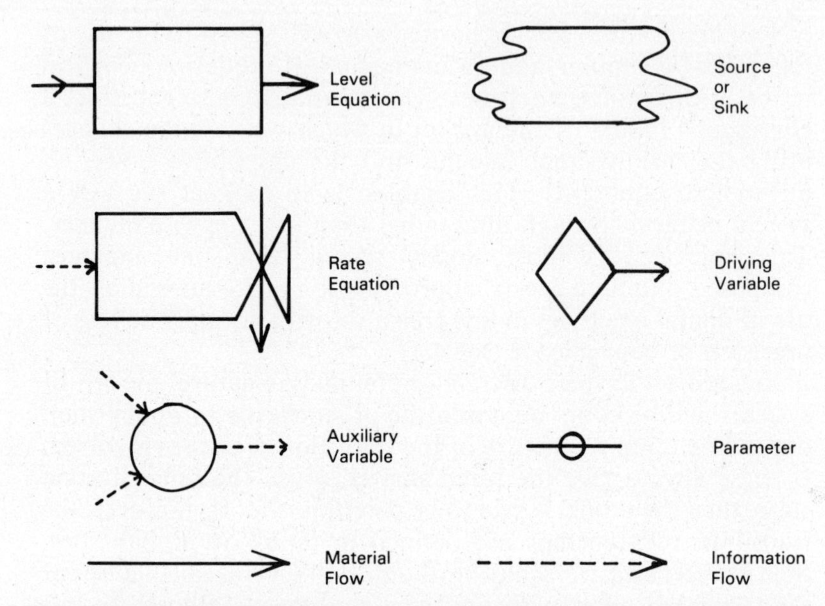

Fig. 7 Some common symbols used in Forrester diagrams. The symbol for "driving variable" is a later addition not due to Forrester. Only flows commonly used in ecological applications are shown

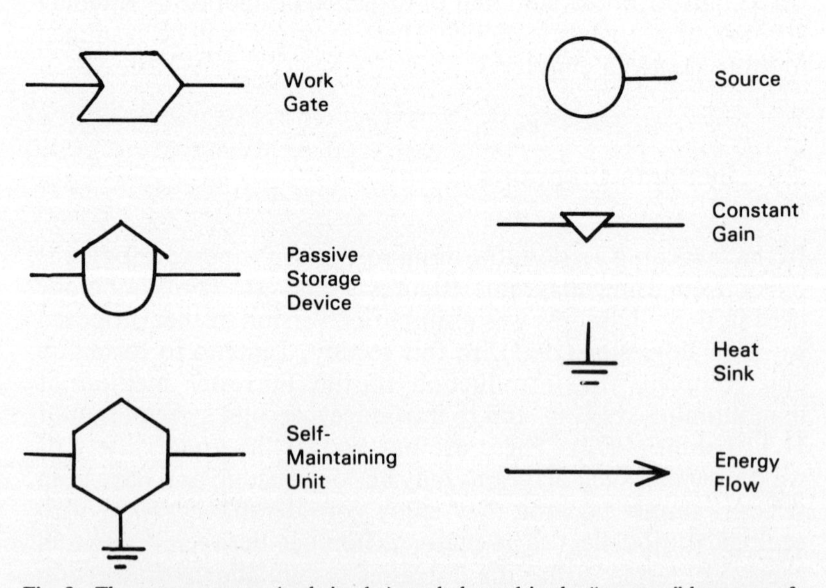

Fig. 8 The more common (and simpler) symbols used in the "energese" language of H.T. Odum

system to another, while the Odum system is strictly used for representing energy flow from place to place in a system.

Figures 7 and 8 show the most common symbols used in each of these two systems. Most of the Forrester symbols shown in figure 7 are already familiar to us, but the addition of two symbols enhances the flexibility of the symbology representing ecological systems:

1. **Auxiliary variables** are represented by circles. These auxiliary variables "collect" information from a variety of sources within the system and deliver it, in summed form, elsewhere. This symbol is useful when a particular collection of information flows can be recognized and named as a separate quantity, such as the level of hunger in a predator or the measure of quality in a pasture and so on. Such a collection of information flows jointly affect some rate equation elsewhere in the system.
2. The other symbol is for a parameter or constant value, and this is useful for representing some known quantity which does not change in response to changes in the system but which nevertheless has an effect on some rate process or auxiliary variable.

In addition to the material and information flows already noted, Forrester (1968) defines a number of other symbols for the representation of flows of money, people, equipment and so on which seldom enter into an ecological system. Figure 8 shows the symbols in the Odum system comparable with those in the Forrester form, plus some other quantities representing ecological units which are not simply represented by single symbols in the Forrester system. The **work-gate** is essentially similar to the rate equation symbol, and the **passive storage device** is also more or less equivalent to the state variable symbol in the Forrester system. Sources of material represented in the Forrester system by irregular shaped boxes are represented by Odum as circles, and the flow of energy around the system is represented by a simple solid arrow. In addition, however, the hexagonal symbol for a **self-maintaining unit** is permissible in Odum diagrams and represents such things as biological entities, individuals or populations, which have an existence independent of the rest of the system. These, nevertheless, process some of the energy passing through the system with concomitant heat loss, represented here and elsewhere in the Odum system by the grounding symbol representative of a heat sink. One other commonly used symbol is the inverted triangle representing a constant gain in electronic terms, which Odum employs when

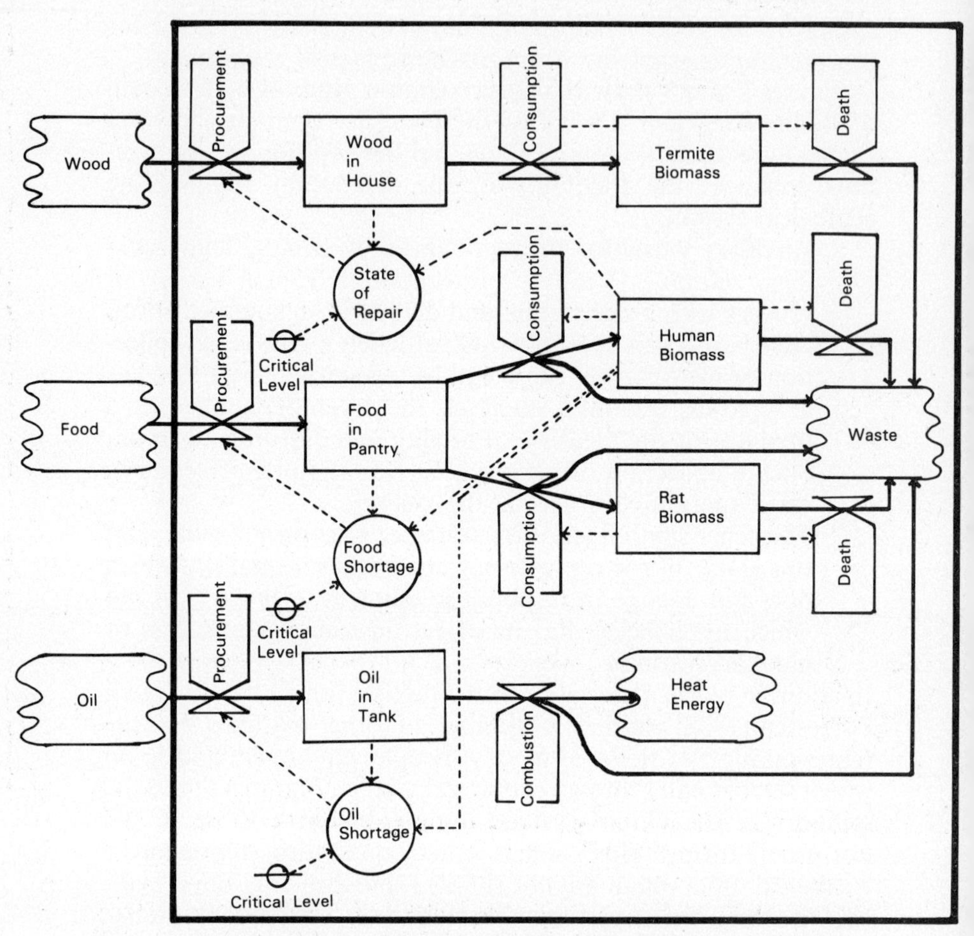

Fig. 9 A simplified domestic ecosystem comprising an oil-heated wooden house with its human inhabitants and the rats and termites that infest it. The system is here represented using the symbology of Forrester, and in figure 10 that of Odum

energy is converted from one form to another in some predetermined and invariant fashion.

The application of these various symbols is perhaps best understood by example: figure 9 shows a simple domestic system using the Forrester diagram to represent material and information flows, and figure 10 using the Odum system representing energy flows. The example is borrowed from Odum's 1971 book. Basically, he defined a house built of wood, containing people and their associated pests, termites and rats, feeding on food brought into the house which in turn was heated by oil also brought into the house. In both systems the stores of

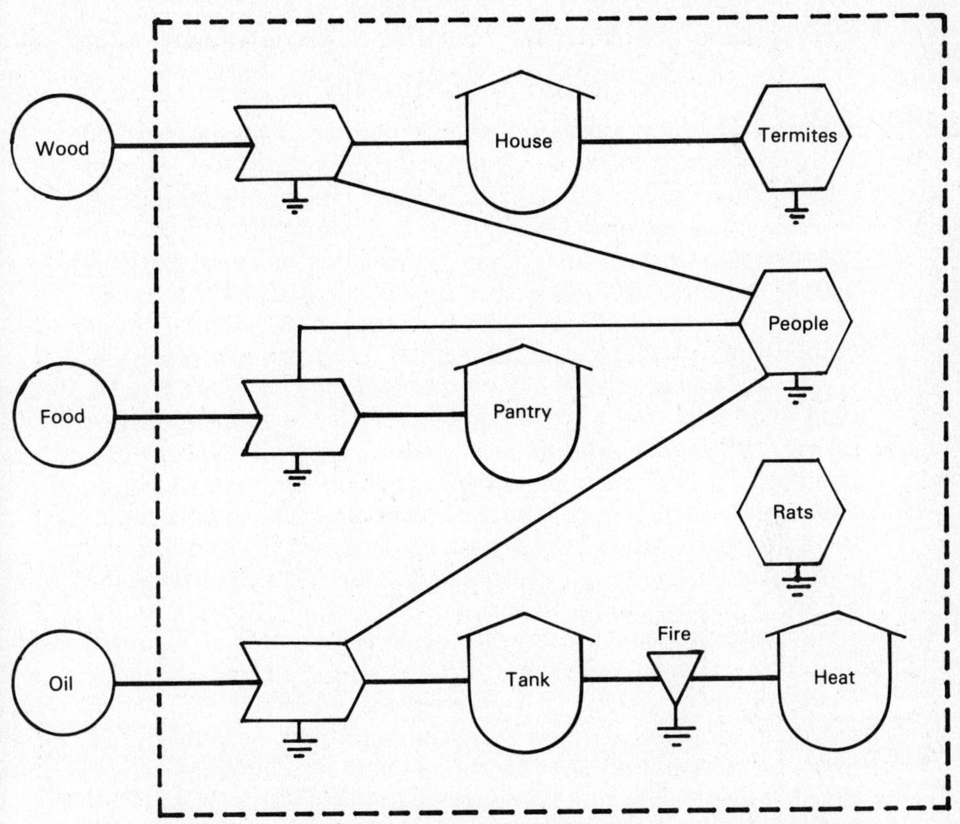

Fig. 10 Domestic ecosystem as in figure 9, represented using the symbology of H.T. Odum, redrawn from Odum (1971)

wood, food and oil within the house are the principal systems variables — represented in the Forrester diagram by the symbols for level equations and in the Odum diagram by the symbols for passive storage devices. These stores are supplied from outside sources of wood, food and oil respectively, represented as occurring outside the system using the appropriate symbols in the two languages. The importation of these materials from outside the system into the house is a rate process in Forrester terms, representing the procurement of these materials, and a work-gate in the Odum system, representing the energy expenditure involved in this procurement procedure. The biological components of the system, the termites, the people and the rats, are very simply represented in the Odum system by the use of the symbol for self-replicating units. In the Forrester system, however, they are represented by additional symbols. These reflect the biomass of each component, which is connected to the stores of materials in the house by more symbols for rate processes representing the consumption of material. The waste products associated with the consumption and death of each of the biological components are represented in material terms, using Forrester's system, by flows to a pool of waste, connected by rate equations governing the death process in the one instance and issuing from the consumption process as waste in the other. In the Odum system, this waste of energy is simply indicated as a heat sink appended to each of the self-replicating units representing these components. The combustion of oil and the subsequent release of heat energy is similarly represented in the two systems. In the Forrester system, the sink of heat energy within the house is fed from the supply of oil in the house via a rate process symbol representing combustion; this also produces a certain amount of waste which, as before, joins the pool of waste that is identified within the system. In the Odum system the heat within the house is represented using a passive storage device which is fed from the oil store symbol via a "constant gain" symbol, representing combustion by fire, waste again being represented as a simple heat sink. These features are those common to both systems. However, examination of figures 9 and 10 shows that much more can be included in the Forrester diagram; we can introduce flows of information not present in the Odum diagram, where the only feedback is that from people to the work-gates involved in the procurement of materials. In the Forrester system, a great many information feedbacks are possible, and for illustrative purposes I have also introduced three auxiliary variables representing the perceived state of the three stores of material within the house. Each of these is fed by information flows from a number of components elsewhere in the system and from constant parameter values representing critical levels in each instance. These auxiliary variables feed into the procurement process.

This simple comparison of the two symbologies may be summarized as follows:

1. The Forrester diagrams represent material and information flows whereas the Odum diagrams represent energy flows only.
2. The Forrester diagrams are more explicit but accordingly more complex in appearance, whereas the Odum diagrams, containing some symbols which represent higher level entities, are more succinct.
3. The Forrester system of symbols is specifically designed for

> the description of complex systems in an entirely general manner and is based on a study of a great many such complex systems. The Odum system on the other hand is a direct product of working with analogue computers and reflects in many ways an electrical engineer's approach to the problem of system representation.

Because of its electrical analogy, the Odum system is relatively easy to turn into mathematical equations, whereas the Forrester diagrams may be more difficult to convert into a series of mathematical expressions and this conversion may require a number of intermediate steps in place of the direct translation possible using the Odum system.

It is neither possible or desirable to say which of these two systems is "better" because they were designed for different purposes. If one is building a model of energy flow then certainly the Odum system should be given serious consideration, but if one is simply trying to put down on paper more or less vague ideas about a system, as the precursor to a more detailed systems study, then, arguably, the Forrester system is the more powerful of the two tools.

3

The Approach to Modelling

Systems ecologists are frequently asked, by colleagues who do not employ the techniques of computing and systems analysis in their work, exactly how one goes about making a model. There has been a great deal of misconception and misunderstanding about this aspect of the systems ecologist's technique and approach to his subject, and most writers in this area have ignored the deeper questions involved. This chapter presents a general account of the processes involved in modelling, and chapters 4 and 5 give more specific comments on the tools and techniques used by systems ecologists. This chapter sets down the **philosophy** involved in the formulation of an ecological model, and the sequence of steps that are followed frequently by the modellers. It cannot be overemphasized that it is this set of concepts that are the most important aspect of the modeller's art; the actual techniques and tools he uses, although necessary, are of secondary concern in understanding the role and importance of modelling in ecology.

The way in which the ecological modeller or, indeed, any other practising scientist works is summarized in figure 11, and can be explained most succinctly by a series of numbered points corresponding with those on the figure.

1. There exists for every practising scientist a body of experience based upon his past work, formal training, general scientific reading and observations made which are peripheral to previous work. These may have been simply stored away mentally or could conceivably have been published as a purely descriptive note by the subject scientist or others in the past (for example, work of the genres: "A preliminary note on ..."; "A new record of ..."; "Aspects of ..."; "An unexplained [something] from ..."). In addition, from past work a scientist has a working intuition about what is or is not likely as explanation in his field of study and, perhaps more important, what is or is not likely to be a good and productive subject of study.

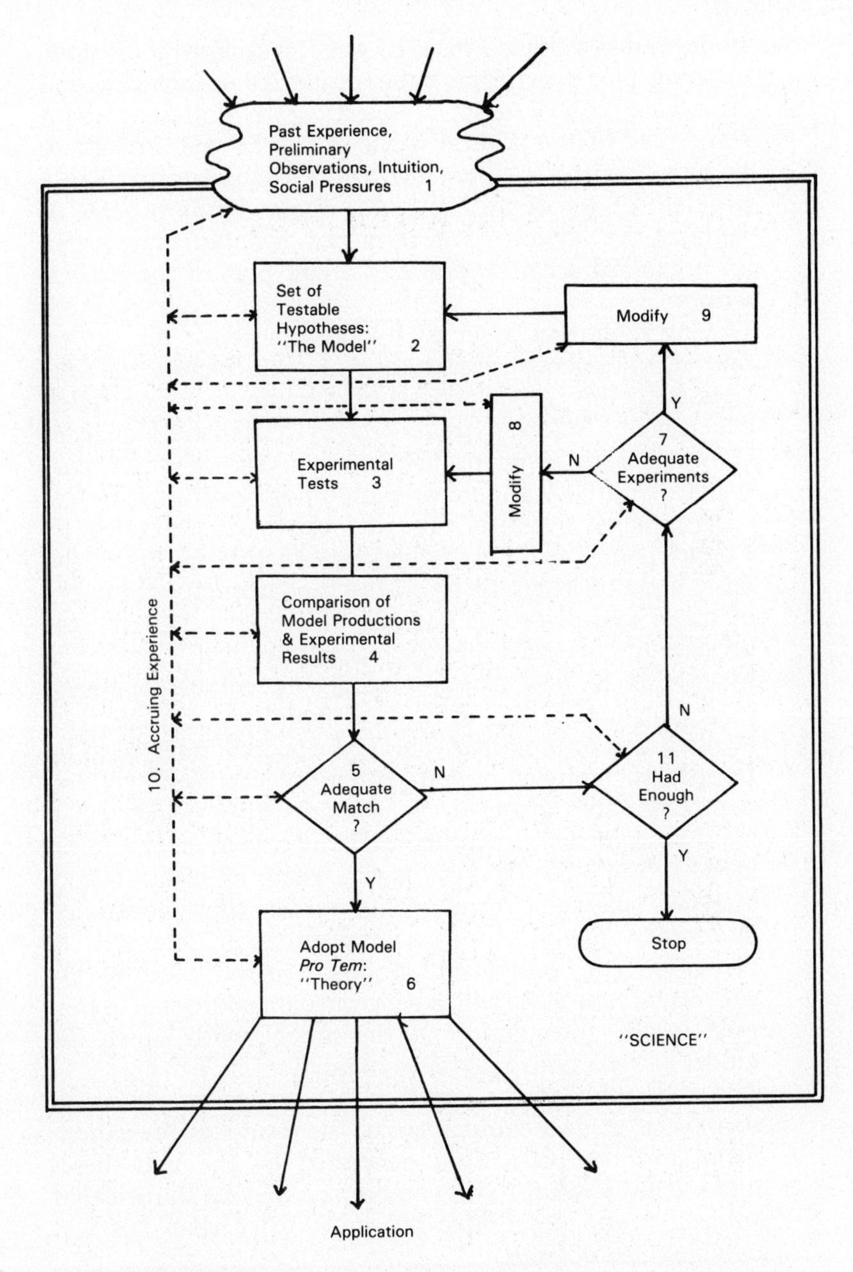

Fig. 11 Schematic representation of the processes of modern science. See text for further explanation

2. In beginning a "new" piece of work, a scientist must draw upon his past experience, either organized or inchoate, and from this, choose a set of questions, hypotheses, about some part of the natural world which he will attempt to answer. For the ecologist taking a systems approach to a problem or set of problems this choice is the process of defining the "system" which, in fact, is nothing more than an organized set of hypotheses about part of the natural world.
3. Having identified a set of hypotheses about his subject matter, the scientist must attempt to put them to the test. Classically this is by a process of a manipulated controlled experiment, followed frequently by a statistical analysis showing, with given probability, whether or not a particular hypothesis should or should not be refuted. In this regard many scientists take the approach described by Popper (1963) as the hypothetico-deductive approach, the basic logic of which turns on the fundamental asymmetry of proof and disproof. Basically this means that whereas no amount of confirmatory information will "prove" a hypothesis to be "true", a single undeniable opposing piece of evidence may disprove a hypothesis, causing it to be discarded or modified. This is the process referred to by Popper as "conjecture and refutation". Accordingly, experiments are arranged as attempts to "disprove" a hypothesis (or set of hypotheses in the case of an ecological model). For the field biologist, such procedures may include "natural" experiments rather than controlled laboratory manipulations. Frequently, in attempts at validation of ecological models, such "natural experiments" are those which generate data additional to that used in the formulation of the model, with which the model's predictions may be compared.
4. The end point of the experimental procedure, whatever form it takes, is a comparison of the results of the experiment and the predictions generated by the hypotheses under consideration. This comparison may be statistical or subjective and descriptive but more often than not is a combination of both.
5. The adequacy of the correspondence between the model's predictions and the set of experimental results is judged using a variety of criteria including statistical "levels of significance". Basically the model must be "good enough" to be applied to the particular problem being tackled.

There are three principle reasons why the correspondence might not be adequate. (a) There may have been some fault with the design of the particular experiment leading to lack of confidence in the results. This may be rectified by redesigning the experiment and repeating the comparison (the (11), (7), (8), (3), (4), (5) cycle in the figure). (b) This route may also be taken when the results of the first experiment are usable, with a suggestive or partial correspondence between what is predicted and observed, but when further, different, experiments are required to give what the scientist considers to be adequate grounds for incorporating the original model in his working theory. (c) Lastly, of course, the experiment or set of experiments may lead to the rejection of the hypothesis wholly or in part, requiring subsequent modification, not just of experimental design but of the hypotheses themselves. This done, the experimental procedure can be begun again (the (11), (7), (9), (2), (3), (4), (5) cycle in the diagram). The two cycles identified may be circumnavigated an indefinite number of times, limited only by personal tenacity, funds or longevity.

6. An acceptable hypothesis, backed by experiments of whatever kind, becomes part of the theoretical base of the science, at least as far as the particular practitioner is concerned. This position of favour, however, is of necessity a precarious one. Even assuming that the reasoning, technique and standards of the original worker are accepted by his peers, any part of scientific theory continually invites examination. It can only ever be a temporary expedient — the best explanation of a phenomenon at any one time.

7,8,9. (See 5.)

10. Peripheral to the various stages of the scientific process as I have described it is a feedback for the scientist or his fellows, into the body of experience and basis for intuition which are the necessary precursors to any further scientific activity. This body of information also provides an input at all stages of the scientific process.

11. To complete the logic of the cyclical processes shown in figure 11, a process must be incorporated whereby a particular scientist can "bale out" of the process due to the complete intractability of the problem or, perhaps, the inabilities of the worker.

The builder of ecological models looks to produce constructs, which have an applicability outside of the data set he used to construct the model. In other words, a model, to be successful, must have a degree of **generality**. This statement, however, is not to be over-interpreted. A particular model, say of a pest population, may be highly successful if it can be transferred and applied to the next field or the next season in the original field. It need not (although, of course, it may) apply across all populations of all pests at all times. It all depends on the sorts of questions the scientist had in mind, or had put to him, orginally. Nor can it be assumed that the more general the model, the better the science — again, it reflects the imposed or self-imposed terms of reference.

This brings us back to the earlier statements about model building, which included the point that successful models must be "good enough" to be applied. "Good enough" is deliberately vague and discretionary. Looking across any of the broad divisions of science, there will be perceived a wide tolerance of what is "good enough". Thus, in a wider context, an error in the model of the ordering of nucleotide bases may make nonsense of a molecular biologist's predictions, whereas an error of a few light years may in no way affect the conclusions of a student of quasars! In some areas of scientific enquiry, the shape of a curve is a "sufficient" result, in others the third decimal place of the slope and intercept parameters may be vital. As before, it comes back to the nature and type of the question or questions being asked.

These general considerations apply across all fields of scientific endeavour, including the whole of ecological science, and the particular techniques of the systems ecologist are no different in basic approach from any other set of techniques which might be applied to understanding the ecology of a population, a process, or an ecosystem. Having said this, it might be pointed out that the advantage of the systems approach in ecology is that a common philosophy and set of techniques can be applied to any ecological system and, in principle at least, this can lead to the identification of generalities across systems, which may not be so apparent when using any of the great diversity of alternative approaches.

What exactly does a modeller do, having made the decision to build a numerical or mathematical representation of an ecological system? Particular workers would of necessity give a very personal answer to this question but, modifying the list of Gordon (1969), an idealized scheme can be examined under a number of headings. In order, an ecological modelling procedure includes some or all of the following stages, which will be discussed in turn.

3.1 Problem Definition
3.2 Systems Identification
3.3 Decisions on Model Type
3.4 Mathematical Formulation
3.5 Decisions on Computing Methods
3.6 Programming
3.7 Parameter Estimation
3.8 Validation
3.9 Experimentation

3.1 Problem Definition

The crucial question that must be asked by someone proposing to build a model of any system is why a numerical or mathematical model is an appropriate way of tackling the problem. It is not sufficient to have a body of data which "looks as though it would be a good basis for a model". This first stage in the model-building process is undoubtedly the most crucial and the one that frequently receives far less attention than its importance warrants. Of course, the definition of the problem may take a great variety of forms. It may be a single question such as, "What will happen if I increase the nutrient input in a woodland ecosystem by 50 per cent?", but, in ecological terms, it is more likely to take a form something like "If I manipulate a woodland ecosystem in any one of a variety of different ways, what changes can I expect to occur in the appearance and dynamics of the ecosystem?" In both cases, a model may be an appropriate way to proceed. For the first sort of question, where there is a specific point at issue, then a relatively simple model may be sufficient to supply the answer required. For the second sort of question, however, at some generally agreed upon level of resolution, what is required is a model of the system concerned which is as accurate and generally useful as possible.

Like biological classification, the modelling exercise is a pragmatic one. There is no point at all in building an ecological model that is more complex, more complete or more time-consuming than is justified by the terms of reference of the problem to which the model is a response. For this reason, the specific problem being tackled must be borne in mind during the model-building process and decisions in all the other more practical aspects of the process made with respect to this.

3.2 Systems Identification

In chapter 1 it was noted that one of the most difficult and important features of any system subject to study was the definition of the systems boundary. The ecological modeller, having decided that the problem being faced demands the construction of a model, is immediately confronted with the necessity of defining the system which he or she is going to study. A choice must be made concerning the level of resolution to be studied — individual organisms, whole populations, perhaps trophic levels, or even larger units within ecosystems. Having made this decision, the ecologist must then select a set of components which interact with each other from among the infinite set of such components available to him from the natural world. There are two different approaches involved here, and the choice depends upon the sort of model intended and so is not entirely independent of 3.3 below. One approach is to make the initial choice of systems components as simple as possible, and to include only those variables which one is confident are critical with respect to the problem one is tackling. Alternatively, it can be argued that the initial model should include all those variables which might conceivably be involved in the processes of interest, happy in the knowledge that subsequent procedures may modify this list of variables until an ideal number is arrived at. The first approach, which may be described as one following an axiom of parsimony, is essential if one intends to build an analytical mathematical model, but is likely to be inappropriate if a complex numerical simulation model is intended. The second approach, on the other hand, which may be described as the generous approach, is wholly inappropriate where an analytical model is intended, but may well be an acceptable approach to a more complex numerical simulation. In practice, the usual route is some middle way in which the components of the system are defined on the basis of previous experience, both as modeller and experimental ecologist.

Once an initial set of variables is identified, it is necessary to start thinking about how they might interact with each other and to identify the nature of these interactions. There are two complementary ways of doing this and both are useful exercises

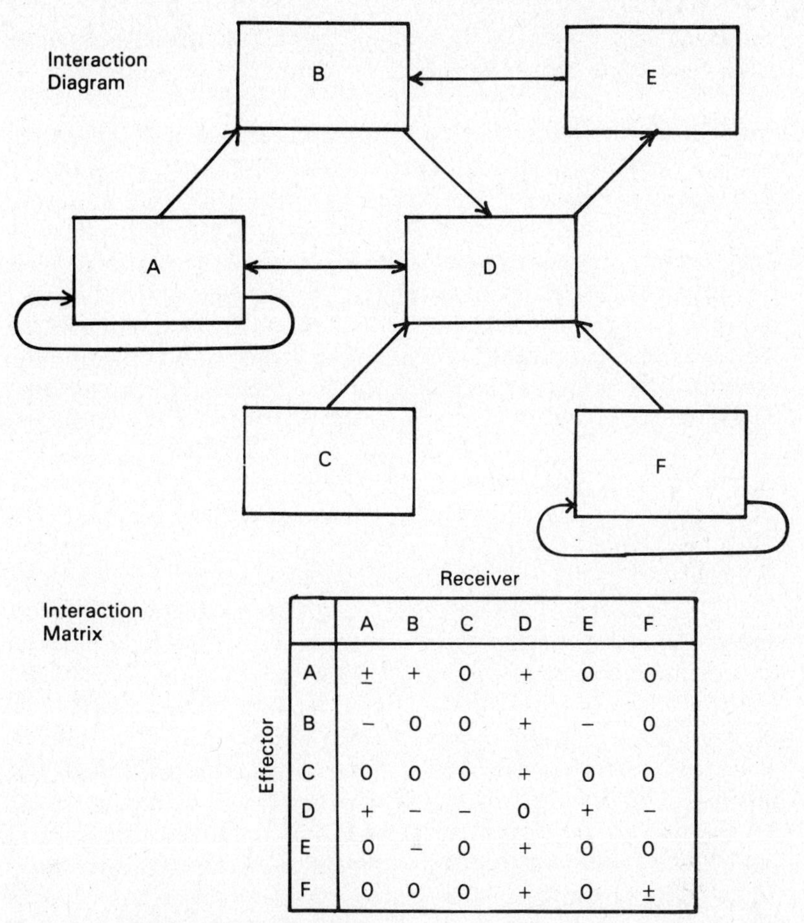

Effector \ Receiver	A	B	C	D	E	F
A	±	+	0	+	0	0
B	−	0	0	+	−	0
C	0	0	0	+	0	0
D	+	−	−	0	+	−
E	0	−	0	+	0	0
F	0	0	0	+	0	±

Fig. 12 A hypothetical system represented by a diagram of interactions and the associated interaction matrix

as precursors to the actual modelling process. The most obvious of these two ways is the construction of a systems diagram, examples of which have been seen in chapter 2. In many instances, the experience of the modeller enables him or her to incorporate interactions between state variables in the diagram without any intermediate step. However, in certain sorts of systems, especially the very complex ones, it may not be possible to ensure that the complete range of interactions has been included simply by going straight to the diagram state. In these instances, a useful intermediate device is the **interaction matrix.** Figure 12 illustrates the use of such a matrix and the connection between the systems diagram and the matrix. Essentially the matrix is constructed by considering each of the state variables within the systems as a source and receiver of signals or flows from each other state variable. The nature of the exchanges between state variables can be positive, that is to say when an increase in the first variable leads to an increase in the level of the second variable; negative, where an increase in the level of the first variable leads to a decrease in the level of the second variable; or, it can have zero value where a change in the level of one variable has no immediate effect on the other variable. It is possible, of course, for a variable to be self-accelerating or self-damping, and in these instances the effect can be incorporated into the interaction matrix by having a positive and negative element on the principal diagonal; often however, these elements would be zero.

3.3 Decisions on Model Type

Once the system of interest has been identified and the interaction between the state variables has been arrived at, it must be decided precisely what kind of model is to be constructed. Many considerations must be taken into account, in particular the purpose for which the model is being constructed.

There are two basic decisions involved: whether the model will be analytical or numerical, and whether it will be stochastic or deterministic. If the system identified is relatively simple, and one can be confident that the interactions between variables are likely to be adequately represented by linear equations without complex time-lags or other mathematical complications, then it may be both feasible and desirable to build an analytical algebraic model. This implies that the model will be constructed using mathematical equations which are analyzable using standard mathematical techniques, and can lead to some elegant structures. This sort of model has been much favoured by mathematicians and physicists entering the field of ecological endeavour. Such activities are well exemplified by some of the recent work of Robert May, whose books, "*Stability and Complexity in Model Ecosystems*" and "*Theoretical Ecology: Principles and Applications*," may be consulted for further information on the approach. A brief introduction to models of this kind is given in 5.1 and 7.1. Frequently, however, ecologists are concerned to make models which are more complex and hence more realistic than can be achieved using analytical techniques. Accordingly, they must use numerical simulation models, which involve more computing methodology than mathematical technique and attempt, basically, to imitate the functioning of an ecological system by use of an appropriately structured computer programme. The computer is used to keep track of the state variables of the model and provide periodic print-outs of the state of the system during the course of any particular run. Such models are sometimes described as "**book-keeping**" structures, although this is something of an over-simplification as almost all numerical simulation models have a set of mathematical relationships at their core, which embody the functional relationships among the state variables of the biological system.

The choice between a deterministic and stochastic approach to modelling is partly linked to the choice between an analytical and a numerical model. Deterministic models, as already explained, are those which are completely invariant in their operation, given an unchanging set of input and parameter values. Most analytical models are of this kind of necessity, although an analytical approach to certain forms of stochastic processes is possible. However, for the non-mathematician, the algebra involved is frequently both esoteric and horrific. On the other hand with a numerical simulation model it is often both convenient and desirable to introduce a stochastic element into its operation. This means that instead of a particular parameter being represented throughout the operation of the model by a constant, a separate value is chosen for each run of the model or even on a number of separate occasions within one run. This method of operation most closely approaches the way we observe particular quantities in the real world, and the set of values from which we choose must be distributed in some fashion which approximates the expected statistical distribution of that particular variable in the real world (see 5.4). There is a variety of different techniques for achieving this aim which will be mentioned later; at this stage it is sufficient to say that they all require a device which can generate "pseudo"-random numbers within the computer. Almost all reasonably sized computer installations provide such a device, as do some of the more advanced hand-calculators, and the actual techniques of simulation involved are correspondingly straight-forward. One distinct advantage of stochastic models for the ecologist is that, when using the model to produce a prediction concerning the consequences of some particular manipulation that might be made in the input values, several runs of the model can be made using different initializing values in the random number generator, thus producing a set of predictions for a particular event which can be summarized statistically. This makes for more direct comparisons with real measurements from experimental or field observations, and may permit the use of standard statistical contingency tests for comparing model output with actual data.

3.4 Mathematical Formulation

Once the nature of the proposed model has been decided, questions must be faced concerning the sort of mathematical structures to be used in constructing the model. A great and ever-increasing variety of techniques is open to the ecologist, but probably that most commonly employed is the use of the infinitesimal calculus, where rate processes which characterize the dynamics of systems of interest are written as **differential equations.** A brief introduction to the nature and mathematics of differential equations is presented in 5.1. Increasingly common alternatives are **difference equations** and **matrix representation**, and these are also dealt with in 5.1 and 5.2. For some purposes the modeller may decide to use more specialized mathematical techniques, which require the model to be couched in a particular fashion. Such techniques include operations research, graph theory, diffusion theory, catastrophe theory and aspects of probability theory such as Markov chains. A treatment of these techniques is beyond the scope of the present work, but some basic entries to the appropriate literature on these and other techniques are laid out in table 1. Jeffers provides a good general introduction to a number of these areas of mathematics in an ecological context in his 1978 book. At this point in the modelling process the ecologist may need specialist knowledge outside his own field. There has been considerable debate as to the effectiveness and feasibility of systems ecologists consulting and using the services of specialists whose primary training is in mathematics (see for example, Gilbert *et al.* 1976, pp. 9-10). Frequently such collaborations result in difficulties in communication. It behoves the ecologist to become familiar with sufficient mathematics either to overcome those problems he encounters in his modelling exercises on his own, or to provide himself with the vocabulary for ready communication with specialist mathematicians. As the obvious counterpart to this requirement, it is highly desirable that the mathematicians involved should familiarize themselves with sufficient biology to understand fully the special biological problems involved. So far, more biologically-trained ecologists have shown themselves willing to venture outside their own speciality and attempt to obtain specialist knowledge in mathematics, computing and other fields necessary to the efficient execution of their modelling activities, than have mathematicians engaged in the reverse flow. There are, of course, notable exceptions to this generalization: mathematicians who have turned to ecology, such as MacArthur and May, have made important contributions in the field.

Table 1. Introductory references to the application of particular mathematical techniques in ecological modelling, other than those dealt with in detail in the text.

Field or Technique	References
Catastrophe Theory	Jones 1977, Jones and Walters 1976, Zeeman 1976, Jeffers 1978
Diffusion Theory	Skellam 1951, 1973
Graph Theory	Gallopin 1972
Control Systems Analysis	Hubbell 1971, Calow 1973
Markov Chains	Horn 1975, Thomson and Vertinsky 1975
Optimization and Operations Research	Peterman 1977
Game Theory	Maynard Smith 1976, Jeffers 1978
Set Theory	Niven 1980

3.5 Decisions on Computing Methods

At this stage in the modelling procedure the system of interest will have been described as a series of mathematical equations representing the dynamics of the selected state variables. In order to proceed to the simulation of the dynamics of the system of interest and its corollary, the assessment of the real value of the model that has been prepared, it is usually necessary to transfer the system of equations into a programme for a high speed computer. A number of decisions about computing activities have to be made at this stage in an attempt to ensure the procedures involved are entirely appropriate in the particular situation.

The first decision relates to the type of machine. Computers are basically of two types, **analogue** and **digital**, which are described in some detail in chapter 4. The choice between the two is usually guided by the purely practical consideration of availability rather than any more objective criteria. If both sorts of machines are available then certain guidelines can be used in making a decision. The reasons for the choices will become clearer when the nature of the two machines has been described later in chapter 4 but, simply, an **analogue** machine is strictly a one-purpose device for mimicking the operation and dynamics of sets of differential equations. It is particularly good at this and the ease and speed of operation make it a real alternative for a model of this kind. If, however, the logic of the model is more complex, and particularly if it involves discrete variables, then the **digital** computer must be used. In fact, most institutions possess only digital computers and these must be adopted for all modelling purposes. Once this decision has been made there is the question of how large a facility is needed. The recent advent of mini- and micro-computers has made this a particularly pertinent question. The obvious advantages of, say, a desk-top machine in terms of accessibility and cheapness must be offset against its limitations in storage capacity or range of programming options. However, accessory storage devices may obviate many size limitations, and micro-computers in general are one of the areas of most technological development and research at the moment. At the other end of the scale, the formation of computer networks via telephone lines means that, in many installations, there is effectively no upper limit to storage space or computational variety, and only cost and scientific prudence limit the modeller's activities.

Once the type and size of machine are decided then programming languages need to be considered. A variety of programming languages can be used for systems simulation and some of these are described in 4.3. The decision once again is constrained by the availability of a particular language on the computer and by the familiarity of the operator with the logic and operation of a particular language. The great majority of ecological models constructed to date have used the near universal FORTRAN language and it can be argued that there is some advantage in favouring this language when any other choice is not absolutely clear cut. Under certain circumstances, some of the special simulation languages that have been developed may be preferred; an introductory account and a consideration of their pros and cons is given in chapter 4.

A second important programming decision concerns the actual method of programming, which depends largely upon the complexity of the system that is to be simulated. The simplest approach to programming is to include all the computations, input and output instructions and internal book-keeping instructions within a single main programme. However, in programmes of only moderate complexity this can lead to confusing and lengthy blocks of code, and major subdivisions, subroutines, functions and so on, may be employed to increase efficiency of operation and interpretation. In recent years a number of higher level methodologies have been developed governing the approach to programming, generally referred to as "structured" programming, and these are described briefly in the next section.

3.6 Programming

Once the language and approach to programming have been decided upon, the systems modeller can proceed to convert the logic of his mathematics into a programme understandable by the computer and, hopefully, by other ecologists at some later stage. The writing of a programme is a very idiosyncratic affair and it is hard to lay down general features of the procedure. A number of stages can be recognized that most programmers pass through in the preparation of any particular job. Firstly the logical flow through the programme must be identified, and this can be achieved most readily using diagrammatic techniques such as tree-diagrams or flowcharts. From these diagrams the programmer can proceed to write out the code involved. This is usually done directly into the computer using a teletype of one sort or another.

The next stage is usually referred to as "debugging", and involves attempting to identify and correct grammatical and syntactical errors in the writing of the programme, and any logical errors which may have crept into the structuring of the code. This is usually carried out in an interactive manner with the computer, which will return error messages indicating the position and nature of programming mistakes and, in some instances, logical errors as part of the operating of the compiler for the particular language being used. Eventually the situation is reached where the programme is acceptable to the computer, that is to say, it does not transgress the rules of the language being used. This, alas, does not necessarily mean that it contains no logical errors, as a set of statements can be perfectly legal FORTRAN, say, but make no sense at all logically, in the same way that one can invent perfectly grammatical but nonsensical English sentences. Under these circumstances, the logic on which the programme was based must be examined again and, if necessary, test data run through the programme to identify the errors concerned. Debugging is an art and there are no hard and fast rules that can be applied. However, a number of techniques are available which can be applied in some circumstances to aid the progress towards a usable programme.

The use of test data and, in addition, the insertion of more output instructions within the programme, in order that intermediate values in the computation become available to the programmer, can help trace the location of logical faults. The more complex the code used the more likelihood there is of logical errors that are hard to trace. When any choice occurs it is preferable to use the simplest and most straightforward code available, even at the expense of a few micro-seconds of computing time. This not only reduces the chance of logical errors and makes debugging procedures simpler but also makes the programme more easily understood by a second user at some time in the future.

One further general point that can be made about programming is that the adequate documentation of any programme is a vital part of the scientific communication process. Such documentation is frequently omitted from programmes written with specific ends in mind, and although good intentions are frequently expressed about adding the documentation later this usually does not happen. Some further comments on the methods of documentation available are given in chapter 4.

3.7 Parameter Estimation

Once the model has been translated into a computer programme which works, it is necessary to return to the original system in order that sensible values may be placed on the various variables and constants which occur in the equations forming the mathematical core of the programme. Indeed, in many cases such parameter estimation is a statistical procedure which, usefully, can be carried on in parallel with the earlier steps in the modelling procedure already described. Sometimes the estimation of a parameter for inclusion in a model is a relatively straightforward process. A body temperature, an altitude or chemical constant can be entered from single observations made in the field or from published values which are well-known and generally accepted. Much more frequently, however, no such values exist and estimation procedures must be entered into before values can be fed into the computer. Often, of course, a single value is not an adequate description of the phenomenon of interest. If a calculation in an ecological simulation model involves rainfall or other weather data, realistically, a different value must be used for every period simulated. This can be obtained from weather records, and a monthly mean based on a number of years' data, or some other summarizing statistic, can be entered into the model as appropriate.

Alternatively, under certain circumstances, raw data for particular periods of time can be used. A common situation encountered with respect to weather records and other ecological data is that no information is available for the particular system of interest and, for a value to be entered into the model, some extrapolation must be made from data relating to an adjacent or parallel system. For biological variables this often implies extracting values from the literature dealing with systems comparable with the one under consideration. As in more conventional statistical exercises, a great deal of care must be taken with any such extrapolation, and the value used must be regarded with circumspection throughout the whole process of simulation and experimentation.

Lastly, it may happen that a parameter has been included in a model where there is no data available on appropriate values. Where this occurs a number of avenues of further progress are possible. Most obviously, one can go and measure the quantities concerned. This, however, may be difficult either because the system of interest may not be readily available at the particular point in time or because the variable identified may be a complex computational one such as "level of hunger" or "predisposition to attack", where the measurement of real values may be difficult or even impossible. Under these circumstances the value used in the model, legitimately, may be a straightout guess. Of course, such a value cannot be regarded and treated in the same way as other parameters in the model. It must be treated tentatively and its role can be evaluated using techniques of sensitivity analysis.

Simply, sensitivity analysis is the evaluation of the relative importance of particular parameters within the model. For instance, it is conceivable that a hundred-fold change in the value of a particular parameter may have only a small effect on the output of the model in any simulation whereas, for another parameter, a very slight change may have significant effects on the output. If the model is very sensitive to a parameter for which there are no data which inspire any confidence, then further work on elucidating the nature and measurement of that variable is vital. If, however, the model is insensitive to variations of substantial size in a particular parameter, then the expenditure of large amounts of time and effort in obtaining an accurate estimate may not be justified. Once again it is difficult to generalize and make rules about these procedures which rest largely on the experience and art of the worker.

A sensitivity analysis of the parameters in an ecological model, testing the effects of a range of values and combinations of values for each, is a highly desirable exercise at this juncture with respect to all parameters, not just the ones with uncertain values. Such a thorough-going sensitivity analysis is one of the best ways of getting a feel for how the model operates, and stands one in good stead for interpreting the results of experimental simulations later in the investigative procedure. In very large models, the numbers of parameters involved make a complete sensitivity analysis impractical. For these cases a variety of summarizing techniques, grouping and/or sampling procedures have been developed, which are discussed in 8.2.

3.8 Validation

The question of the validation of an ecological model, that is an evaluation of how accurate it is in simulating the real system under consideration, is a vexed one. The difficulty lies in the fact that each systems ecologist must decide when the accuracy displayed by a particular model is adequate for his purposes. This does not mean that another worker looking at the same model will agree on the criteria to be used in this decision, and dissention can arise.

Tests must be made to ensure that the model reproduces accurately the data that were used in its construction. This process of verification, as it is sometimes called, is tautologous in principle but nevertheless a necessary check that the mechanisms of the model are in fact doing what the modeller thinks they are doing. In a sense the modelling process is a data- fitting exercise analogous to the least squares technique of statistics, and the verification process is comparable with a test of ''goodness-of-fit''.

It is necessary also to compare the output from the simulation with a set of test data before using the model for any experimental and/or predictive purposes. This can be illustrated with an imaginary example. Let us suppose that we have built a simulation model of the dynamics of an animal population based on a collection of data from a particular site over a three year period. We have written equations describing the dynamics of a number of age classes in the population and have put numerical values on the parameters involved, extracted from the three years' data. As a result of our verification processes, we know that, given the values of the driving variables, such as temperature, availability of food and so on, that were available during that three year period, we can reproduce the dynamics of the population. The test of the model comes when we feed in the values of driving variables collected either in a separate period of time from the same site or over the same period of time from a different site. If our model is a good one it should then predict the dynamics of the animal population at this second site with an acceptable level of accuracy. What we are looking for is a generality in the model. The output from the model can be compared with test data from the real world using straightforward statistical techniques, correlation and regression analyses, and simple tests of significance. It should be pointed out here that, in practice, the first validation procedure applied to any particular model will almost certainly lead to the identification of major differences between the model's predictions and real world data.

There is an argument which has been used in recent years which goes contrary to the validation procedure just outlined. This says that all possible data available at any point in time should be used in parameter estimation so that the model's descriptive power is maximized at that time. It is hard to reconcile this notion with the principle of validation as outlined above, but perhaps a modification of the proposed techniques would be to use test data to validate a model, but then to incorporate that test data in a rerun of the parameter estimation procedures already employed, to improve the accuracy of the figures used. This still means that at some stage a body of information has to be identified as test data, and used in order to give some impression of just how good a copy of the real world the model represents, but it envisages a "working" model which is always one step ahead of the validated model in the accuracy of estimation of its parameters.

3.9 Experimentation

Once all the procedures described above have been executed and the systems ecologist has a model of the system of interest which matches well in its predictions with other available data, then the model can be put to use as a research tool. In other words, the ecologist can attempt to answer those questions which he identified at the very beginning of the modelling process and which, indeed, were the justification for the whole exercise. This is usually described as experimentation, and involves the manipulation of values within the model in order to assess the response of the system to these pertubations. These may range from comparison of different management strategies on population structure, to the evaluation of new chemical treatments in agriculture, to identifying critical values of elements in aquatic systems which may lead to eutrophication and so on. The list of possible problems which may be tackled in this manner is endless. A good simulation model may be an essential research tool in this regard, as large scale manipulations can be simulated and their effects evaluated before any field operation is undertaken. Then the field trial need only be implemented when the model suggests that the results obtained from such manipulation would be of considerable ecological significance, or would produce the desired result in a managed system. It cannot be overemphasized, however, that a prediction by a simulation model is useless by itself; the employment of such models in ecological research must go hand in hand with parallel field and laboratory activities. Policy decisions in management, for instance, can be made only on the basis both of the predictions of the model and the results of on-site trials.

The many and varied ways in which these steps have been put into practice are reviewed with detailed examples in chapters 6 to 8, covering a variety of ecological levels.

4

Computers and Computing

The electronic computer has become the hallmark of systems ecologists. Although it is possible to conceive of building a model of a complex ecological system without the aid of such machines, the prospect is daunting indeed. All would-be systems ecologists, all ecologists in fact, need to know the elementary principles of operation of the computer and how to use the machine to their advantage — to make it do what they want rather than tailoring their work to the idiosyncracies of whatever computers are on hand. This section provides the basic information required on the functioning and use of the computer, although the student must turn to other works if he wishes, say, to learn a particular programming language; reference is made to appropriate manuals below. Useful introductions to the fascinating field of computer science are provided by Davis (1965), Bartee (1960) and Ralston (1971).

There are three general points, often overlooked in the popular view of computers, which must be borne in mind in developing any familiarity with the machines at all:

1. Computers will do precisely what they are told to do and nothing more. One cannot assume that the machine can make the logical jumps which are an intrinsic part of human thinking. For example, if we wish to add the numbers one to ten mentally we simply tot them up in our minds and come up with the answer. However, we have made several assumptions that the computer, set the same task, cannot make. We have assumed not only that there is a variable, the running total, but that this quantity is zero before we start. In addition, when we come to add the last number, we stop because there are no more numbers in the sequence of interest; we do not need to count the number of figures in the sequence to know how many additions to carry out. In programming the machine to do this arithmetic, we must be explicit about all of these things: we set the running total to zero initially, specify the number of

Table 2. FORTRAN programme for the addition of the numbers 1 to 10

```
        PROGRAM INTSUM
        ISUM = 0[1]
        INT = 0[2]
        DO 100 J = 1,10[3]
        INT = INT + 1[4]
        ISUM = ISUM + INT[5]
100     CONTINUE
        WRITE(50,200)ISUM[6]
200     FORMAT(1H0,30HSUM OF THE NUMBERS 1 TO 10 IS 4)[7]
        STOP[8]
        END
```

Note: Superscripts are NOT part of the programme.

1. Sets the running total to zero.
2. Sets the integer to be added to zero.
3. A LOOP instruction telling the machine to perform all the operations up to and including that labelled 100, ten times.
4. Increments the integer to be added by 1.
5. Adds the integer to the running total and then returns to the DO statement (from the subsequent CONTINUE instruction).
6. Instructs the machine to output the final sum on the device labelled "50" according to the format given in statement 200.
7. Gives the format for writing out of the final sum, that is, on a new page of the printer (the 0 part of 1H0) write the words "Sum of the numbers 1 to 10 is", followed by the sum which may be a four digit integer ("I4").
8. Tells the machine to stop.

additions, tell the machine what to do with the answer and how to do it. This is exemplified as a FORTRAN programme with explanatory notes in table 2.

2. The computer's strongest point is that it can perform innumerable complex sequences of arithmetic with very high and repeatable accuracy. In this way it can handle large quantities of data or long runs of repetitive calculations which would defeat hand operation by their sheer magnitude — it is a high grade "number cruncher" to use a commonly applied epithet.
3. Most importantly, the computer is a *tool* to be used by the systems ecologist, and must not become his master. Too frequently new schisms in ecology have come about by the elevation of a particular piece of technology to a central position in a school of thought. This is as large a danger in the case of the computer; in many instances workers have been entranced by the elegance of modern computer technology to the point where their main preoccupation is with using the most sophisticated devices available, producing the most parsimonious (and usually in consequence, most obscure) programmes, or seeking out problems which

permit the widest and technologically most interesting use of their computer installation. This is often to the detriment of their ecological activities and is surely to be avoided.

4.1 The Machinery

4.1.1 Analogue Computers

In most installations the world "computer" implies the digital variety which will be discussed at greater length in 4.1.2. However, the other basic variety of machine, the analogue computer, must not be overlooked in considering ecological applications. Indeed, one very prominent school of systems ecology, led by H.T Odum, has developed around the use of analogue machines and an example of a piece of work from this group is described in 8.3.

As its name implies, the analogue computer provides a copy (the analogue) of the dynamics of the system of interest using a different medium. The changes and flows analogous to the numerical ones observed in the system of interest are electrical. In essence, the analogue computer mimics the operation of differential equations (see chapter 5), using different pieces of electrical circuitry to modify current flow through the machine. The flows of electricity are of course continuous, and this permits the achievement of a very close correspondence between the model system and the mathematical equations being used. The machine itself contains a large number of resistors, potentiometers, comparators, capacitors and amplifiers which can be connected (or "patched") together with wires to produce whatever circuitry is required. The most prominent visual features of an analogue computer are, in fact, the patch boards on its front surface, reminiscent of an old-fashioned, plug and socket telephone exchange. Four basic mathematical processes can be achieved in this way: summation, integration, multiplication and arbitrary function generation (Patten 1971). The use of analogue computers is restricted to the implementation of differential equation models but, accepting this limitation, they do provide a tool on which gaming with the parameters of a model is both easy and well-nigh instantaneous. The effect of increasing the growth factor in a population model, for instance, can be observed simply by twisting the knob of a potentiometer and following the change in output signal on a monitor screen. This facility is not available on digital computers, although many have "analogue simulators" such as IBM's simulation package, CSMP (Continuous System Modelling Programme), which permit the use of digital machines in an interactive fashion as if they were analogue computers. However, even the best of these is more cumbersome in operation for these restricted purposes than an analogue machine.

Patten (1971) provides an introduction to analogue computing for ecologists and Hargrave (1972) provides a more technical approach to the topic.

4.1.2 Digital Computers

Whereas analogue computers are continuous devices, digital computers operate only on discrete numbers — albeit of such length as to approximate the continuous situation. The basic structure of the digital computer is illustrated in figure 13.

The central features of any digital computer installation are the storage devices. These can be divided into two main sorts, the core memory and the accessory storage devices. The core memory is an integral part of the digital computer and is the

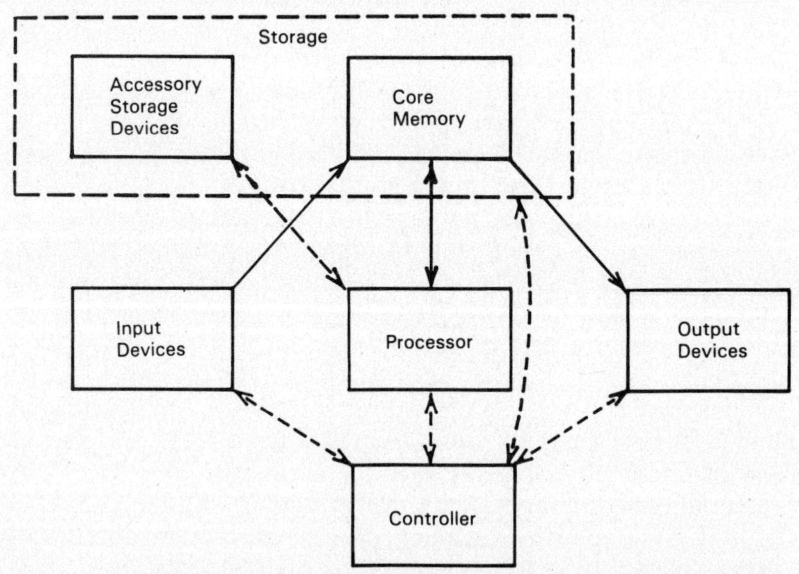

Fig. 13 The basic structure of a digital computer. Solid lines represent flows of data through the machine, dashed lines represent pathways of controlling instructions within it

place where data and programmes, which are being operated on by the machine at any point in time, are stored. All information in the core of a digital computer is stored in binary form. This can best be understood by realizing that the numbers that we normally use are, in fact, a specialized subset of the universal set of numbers, and are characterized by the fact that they are based on a standard unit size of 10. They are numbers to the base 10. To represent them properly one must have ten different symbols, 0 through 9. Binary numbers are simply numbers to the base 2 and for their representation need only two symbols, 0 and 1. This can then be implemented readily electronically because the direction of electrical charge on a wire, for example, can be taken as equivalent to either 0 or 1 as the case may be. Numbers are readily converted from one base to the other and table 3 shows how the simple conversion from base 10 to base 2 can be made. In addition, the table shows a few examples of binary numbers with their base 10 equivalents. It should be

Table 3. Binary representation of numbers

Any ordinary (base 10) number can be converted to binary (base 2) representation by dividing the number repeatedly by 2 until the answer is 1. The final answer and the remainders along the way give the binary equivalent.
Example: To convert 333 to a binary number

Division	*Remainders*
2 | 333	
2 | 166	1
2 | 83	0
2 | 41	1
2 | 20	1
2 | 10	0
2 | 5	0
2 | 2	1
2 | 1	1
0	

Reading from the bottom we obtain the binary number 11001101

i.e. $333_{10} \equiv 11001101_2$

Some other equivalents

$$1_{10} \equiv 1$$
$$2_{10} \equiv 10$$
$$10_{10} \equiv 1010$$
$$100_{10} \equiv 1100100$$
$$1000_{10} \equiv 1111101000$$

remembered that the memory of a computer contains all information relevant to the operation of the computer: the programme which tells the machine what to do with data, the operating system which tells the machine itself how to function, and the data on which the machine is to operate. In addition to the core memory, a large number of different accessory devices are available in modern computer installations, on which programmes and data can be stored when they are not required by the computer for processing. These range from now-outdated punched cards and tapes, through magnetic tapes and drums, to devices such as hard and "floppy" discs. These devices will be dealt with in more detail below. Before information stored on these devices can be used by the computer, however, it must be transferred into the core memory.

The processing element in a digital computer is the one which can most readily be compared with a calculating machine. This device contains the electronic circuitry which enables the basic arithmetic processes of addition, subtraction, multiplication and division to be carried out together with a few more sophisticated transformations. The various components that carry out these arithmetic operations are constructed using binary logic. This is comparable to the normal logic used in base 10 arithmetic, but of course handles only the symbols 0 and 1. The algebra associated with this sort of operation is referred to as Boolean, and the basic laws of Boolean algebra, the building of circuits using these laws and the ways in which this information is applied in computer construction is described in some detail in Bartee (1960). The basic mode of operation of the processor is that a number or set of numbers is received from the core memory, some operation is carried out upon it and the result is returned to the memory element.

The input and output devices are perhaps the most important part of the computer from the user's point of view. The input devices may be punched card readers, paper tape readers, magnetic tape readers, as well as teletypes and other forms of manual direct entry to the machine. Information entered into the computer by one of these devices is stored in an appropriate location in the core memory, where it can remain to be operated upon as the needs of any particular computation demand. Output devices receive the results of computations from the memory unit on request and translate them into some format acceptable by the user. A great variety of output devices is available in modern installations: among the most common are printers, cathode ray tubes and graphical plotters. Input and output devices are described in more detail in 4.1.3.

The last component of the basic computer illustrated in figure 13 is perhaps the most important to the computer, as opposed to the user. This is the controller. Every computer operation requires that information be transferred from an input device or the memory to another location in the memory perhaps, or to the processor, or the output device in use, and so on. The sequence of operations and the way in which they are coordinated is one of the most critical features of the computer design. Within the computer itself the control unit directs the operation of all the other components. In computation it interprets the programme and organizes the rest of the machine to comply with the instruction.

There are a great many variations on the basic design in modern computer installations, but the components identified and described above are the essential units involved. In large installations several of these may be duplicated, and even in relatively modest installations one would expect to find a variety of input, output and storage devices.

4.1.3 Input, Output and Storage Devices

Although the operation of electronic computers depends more upon the nature and efficiency of the controller and memory units, the user is more likely to come into contact with one or more of the various peripheral devices associated with the installation. Indeed, in modern installations the number and variety of these devices is frequently of more importance in determining the range of problems that can be solved than the more operationally central computational units. In addition, it is the transfer of information between these devices and the central processor that determines the speed of operation of the whole installation. The most sure way of increasing the efficiency and usefulness of a computer is not, as is frequently imagined, to add another processor or another memory unit, but to increase the number and variety of input, output and peripheral storage devices.

Most users are introduced to data preparation and programming through the use of keyboard machines such as teletypes, which produce a record of what is typed on paper, or visual display units in which the record appears on a television screen. These devices have replaced almost completely the card-punches which provided the principal means of input to the computer until a few years ago.

As with typewriters, the arrangement of symbols and letters on the keyboard of these machines is standard, and skilled operators are able to process information at a very high rate. Unlike punched cards and tapes, keyboard terminals provide direct access to the computer. Instead of information being typed onto cards or tape, the lines of code or data go directly into a storage location in the memory of the computer itself. These can be transferred later to a less centrally-situated storage device. Such devices are particularly useful in the editing of programmes, whereby a programme or set of data held in the machine can be recalled line by line if necessary, and changed by a process of correction and erasure using the local keyboard; the corrected or altered line is sent back to the central store of the computer to replace the particular set of code in its original sequence. Normally, of course, these direct-access devices are situated close to the computer installation itself, but the development of the technology of telephone links has permitted them to be considerably remote from the central machine in many instances. Special telephone lines can connect a remote terminal or node to the central machine, or the terminal may actually connect through a regular telephone to the computer using public telephone lines.

Modern computers are able to receive input from many different sources simultaneously as far as the user is concerned. They do this by various devices involving swapping programmes in and out of the central processor as required, and keeping track of the current instruction for each of many operations at the same time. Much of the modern explosion in computer technology has been involved in the development of this sort of multiple access facility. A newer development still is the linking of many computers via telephone lines, and the whole supercomputer thus created is used in a multiple access fashion, with many hundreds or even thousands of users at any particular time.

For ecologists wishing to analyze spatial patterns, digitizers are particularly useful. These are devices rather like map-tables in which a cursor of some sort, a sound pen, light pen or other electro-magnetic signal generator, is moved about a map, a tracing or other two-dimensional representation of pattern, and enters particular coordinates at the push of a button into the memory of the computer or some peripheral storage device.

Other forms of input to the computer are most commonly

associated with the transfer of data from some remote storage device into the central processing unit of the machine; these will be dealt with in the account of storage devices below.

Other less common input devices include readers which can cope with magnetic ink, most often used in such applications as the computerization of bank cheques, the use of mark-sense cards where punched cards are replaced by a printed card on which information is coded using a pencil mark, and various other forms of optical recognition by the machine which are used in repetitive situations such as invoice processing, and other business applications.

Voice recognition and the output corollary, voice synthesizing devices, are now in existence and will presumably become readily available in the near future.

Output from the machine is similarly diverse, and many of the devices available complement corresponding input devices. The most common output device, however, is still the printer. These devices will be known to most people who have had anything at all to do with computers; they comprise a very fast printing device producing lines of print on the familiar wide computer paper, which has 132 printing positions in most installations. The positioning instructions with respect to output, that is to say whether or not a particular line is a new line or a new page, are controlled from within the programme being processed in the machine, and special carriage control information must be included in format statements (see table 2). Output can also be obtained as punched tape or on other forms of peripheral storage devices such as magnetic tapes, discs and drums (see below). Output to direct access terminals either involving printers or cathode ray displays has already been mentioned; these have the same computing restrictions as input using the same devices.

One particularly useful form of output available to the simulation modeller is the on-line graph plotter. A variety of such devices are now available, of different sizes and technical capabilities. The systems ecologist can make very good use of even a twelve-inch plotter of this sort and, used with care, it can simplify many of the problems associated with the display and presentation of the results of simulation models. Such plotters usually operate in response to a special routine written into the programme, which manipulates data, simulates processes and so on. It is necessary in such a programme to specify the scales on both abcissa and ordinate and the range of numbers on such scales. The more advanced plotters have a large number of built-in instructions, however, and their use is accordingly very simple. The most obvious and common use of such machines is to produce two-dimensional graphs in the normal way, and frequently the results of simulations can be reduced to a series of such graphs. With a little additional programming, they can be made to produce three-dimensional plots, spatial diagrams of various sorts, maps, dispersion diagrams and so on. Some of the results of such plotting are shown in chapter 7. A great variety of materials can be used on a plotter from temporary ball-point plots on paper, to permanent Indian ink plots in a variety of colours, to plots on transparent film for use in projectors. The standard of production of some of the more advanced machines is such that plots can be used for publication purposes directly. An alternative to using the actual plotter is frequently available in the form of a cathode ray screen, permitting the plots or other diagrams required to be examined on the temporary medium of the screen before permanent copy is made using the plotter. Some workers have gone overboard in the use of plotters. Although there is absolutely no doubt that plotters are useful and in some instances essential adjuncts to any computing facility, they can be made to produce very complex diagrams, three-dimensional figures and even pin-up calendars and, as with several items of computing machinery, the temptation is to become so engrossed in the capabilities of the facility that the scientific considerations involved become neglected.

Several forms of storage device have already been mentioned. Magnetic discs and tapes are the most commonly used devices for programme or data storage. Large machines still favour "hard" magnetic discs on which information is recorded as on a gramophone record. Such discs usually come in stacks of from five to a hundred, such a stack being referred to as a "disc-pack", and information is stored on both surfaces of each disc. It is therefore a very compact form of data storage and has the advantage over tapes that any segment of any of the discs in a pack can be accessed directly, that is to say it is not necessary to run through a great deal of unwanted information in order to find the segment that you are interested in. This is achieved by having a series of read/write heads on movable arms as part of the disc reader in the computer installation, and these, suitably programmed, can seek out particular segments of stored material on discs. Hard discs are used for more permanent storage. Flexible "floppy" discs and magnetic tape are favoured for more transient storage and are also much used in smaller installations, including microcomputers (see 4.1.4). Floppy discs are inexpensive but fragile devices which are the main storage medium for microcomputers, although they are used to some extent in larger machines. Magnetic tapes in cassettes provide similarly inexpensive information storage and remain the most portable of the range of available devices.

Various other storage devices, magnetic drums, punched cards and paper tapes may still be encountered from time to time, but they are outmoded.

The two basic reasons for seeking external storage devices — portability and for avoiding the expense of core storage — have both been undermined by recent technological developments. The flowering of semi-conductor technology has led to the production of a form of computer memory which is cheap both to build and maintain, and is compact and portable. This sort of development, already in production, points the way to new developments in data handling. It will replace the peripheral storage devices discussed above and may even lead to an entirely new concept in computer memory, where the semi-conductor storage units themselves make up the equivalent of the core memory. They are modular and can be carried from one installation to the next, they can be linked together to form more and more capacity for a particular machine. Such devices may have particular programmes built into their circuitry ("Read-only memories" or ROMs), or can be used for casual information storage just as can discs ("Random access memories" or RAMs). The disadvantage in this latter case is that a continuous input of power is needed to maintain information in the memory — turn off the computer and you lose your programme or data.

Even more recent in the evolution of storage media is the so-called "bubble" devices which circumvent the need for continuous power input, based as they are on patterns of deformation within inorganic crystals. Such devices are in the prototype stage at the time of writing in 1982, but almost certainly will appear on the market in the near future.

Levine (1982) describes the development of so-called supercomputers — machines of mind-bending complexity, speed and computational ability exploiting recent advances in miniaturization. Machines in existence in 1982 have achieved average rates of operation of twenty million calculations each second. Levine demonstrates the usefulness of such vast machines in engineering design and evaluation, and even in the design of further generations of supercomputers. No ecological applications of such, as yet, experimental machines are known to me, although they are to be employed in the related field of meteorological forecasting. Undoubtedly they have the computational power to handle the most complex problems of environmental management or ecosystem dynamical prediction. A more apposite question is whether we, the ecologists, have the wit yet to programme them to do so!

4.1.4 Microcomputers

A new variety of computer has recently become commonplace and much used in ecological work. These are the so-called microcomputers, which have the same basic architecture as the standard large, digital computers illustrated in figure 13 but on a smaller scale. Operated individually they act as a sort of super-calculator for the user. However, their capabilities and usefulness should not be underestimated: microcomputers are available in a wide variety of designs and sizes spanning the range from the calculator to the "normal" computer.

The information storage devices most commonly used on microcomputers are floppy discs but hard discs, magnetic tape and even bubble storage are available on some machines. Input and output devices have the same range as for larger computers and include, commonly, visual display units, graphics terminals and keyboard printers. Voice recognition and synthesis devices are also available in some cases.

Most microcomputers operate using BASIC language (see 4.2) but compilers in FORTRAN and other high level languages are available on the more sophisticated "micros".

A wide variety of applications is reported for microcomputers and, for the ecologist, they offer great opportunities for interactive modelling of simpler population and individual-based systems. Nevertheless, large scale, realistic simulations of life systems and ecosystems remain the province of larger machines. Webster (1981/82) and Spain (1982) provide introductions to microcomputers and their applications in biology.

4.2 Programming

4.2.1 Some Introductory Concepts

The actual operation by the computer is by the manipulation of binary numbers. In order to speak to the computer, therefore, the immediate requirement is to use binary numbers, which is more or less what early computer users had to do. They had to prepare their instructions for the machine in machine code which was specific in its structure and rules to particular machines and, indeed, to particular installations. This is a cumbersome procedure and does not lend itself very well to particular applications. It is slow and wasteful in so far as somebody operating on one machine must relearn his method of communication on moving to another machine. Machine code represents the operating language of a particular computer. When we wish to communicate with modern computers we generally express our instructions in a language which is easier for us to understand and which the machine can understand and convert into its own machine code for implementation. Such higher level languages allow us to express the logic of the programme we have designed in a form which the computer can understand. An analogy used by Ralston (1971) is that, as with two people whose native language differs, one of the basic ways of communicating is for both to speak a comon third language. This is essentially what we are doing in using higher level languages to programme the computer. Its language, machine code, is very hard for us to understand and our language, English or mathematics, say, is incomprehensible to the computer. However, we can both use a higher level language such as FORTRAN or COBOL.

There are four principal higher level languages in common use at present: FORTRAN, ALGOL, PASCAL and COBOL. Manuals for these languages are McCracken (1972) – FORTRAN, McCracken (1962) – ALGOL, Ashley (1974) – COBOL, and Jensen and Wirth (1975) – PASCAL. FORTRAN and ALGOL are algebraic languages in which most scientific work is carried out. COBOL was designed as a business language and is particularly apt for file maintenance, inventory control and similar activities. PASCAL is a relatively new language which provides facilities especially appropriate for the production of structured programmes (see 4.2.3). It has been developed from the ALGOL language but has much extended data-structuring options. Other languages are available and some are commonly used as training tools in introducing people to computer programming. The most commonly used of those not so far mentioned, BASIC, is closely related to a simplified form of FORTRAN and is probably the principal language used on microcomputers (see 4.1.4).

These higher level languages are converted into machine code through an appropriate compiler. A compiler is part of the computer's operating system and is usually machine specific. That is, each major computer manufacturer produces a FORTRAN compiler, a COBOL compiler and so on. For a variety of reasons, each manufacturer also builds in extra features to this compiler which are not necessarily present in other compilers of the same language. This has led to a distinction between installation-specific language and standard language for each of the major higher level languages. It cannot be stressed too heavily that, wherever possible, the internationally accepted standard form of particular languages should be used, even if this is slightly more cumbersome in a few instances. Only if this sort of standardization is achieved can programmes be exchanged between computers, and the true universality of the approach exploited.

So far two strata of languages have been identified in computing, the machine level of language and the higher level language. There is a third stratum sometimes referred to as simulation languages. These structures, of which a great variety are available, have been used by some workers and provide short cuts to many of the commonly used procedures in higher level languages. Thus a graphing procedure or a sorting procedure or a tabling procedure requires a considerable block of code in most of the more usual higher level languages. What a simulation language may do is to build into its compiler such blocks of code and replace them for the user by a single instruction. The distinction between higher level languages and simulation languages, however, is not a fundamental one, as most available simulation languages are based on particular higher level languages. Thus SIMSCRIPT has its basis in the ALGOL language, GASP is FORTRAN based and so on. Such languages are very useful when a sustained sequence of simulations is to be carried out within a particular project; frequently each has features which are particularly suitable, for instance, in the production of models in which there is a great deal of data handling and presentation. Offset against these advantages, however, are the facts that there are few universally available simulation languages, and the knowledge of such languages among users is somewhat restricted.

The systems ecologist or other computer user approaching the idea of programming for the first time frequently asks what language he should learn. It is not simple to give a straightforward answer to such a question. FORTRAN is the most commonly used and available of all computer languages, which lends weight to the argument that, for maximum transportability and acceptability, simulation models built by ecologists should be written in FORTRAN. Offset against this, however, is the argument that certain forms of operation common within simulation procedures are not easily implemented in FORTRAN, and there is a case for the use of simulation languages or other higher level languages such as PASCAL for these purposes. In addition, of course, the conservative argument that we should continue to do what we have always done in the past is not one generally conducive to scientific or any other sort of progress. Probably the most honest and helpful answer that can be given is that one should learn the principles of computing, its notions of logic and structuring, using one higher level language but paying attention to the general features of the programming procedure rather than becoming particularly attached to the grammar and syntax of the specific language being used. Indeed, an experienced programmer has little difficulty in picking up a programming language unknown to him once he has grasped the basic elements of programming with respect to a particular language, whatever that may be.

4.2.2 Programming Languages

In this section, a comparison will be made between three of the most commonly used higher level languages, namely FORTRAN, ALGOL and COBOL. FORTRAN and ALGOL are most often used in scientific work, but for the fuller understanding of programming systems it is instructive to include a brief look at a language designed for a different purpose, namely COBOL. Tables 4, 5 and 6 show the programme appropriate for the solving of a particular problem written in each of the three languages concerned. I have chosen a somewhat trivial example for this exercise and have written programmes which read a sequence of ten numbers, add these ten numbers, calculate the mean and determine whether or not this mean is more than one hundred. Each programme prints out a message indicating whether or not the mean is within bounds or out of bounds. I have in each case attempted to keep the programme as simple as possible, and have omitted such things as headings for outputs, the writing out of data as it is read in and comments on what the programme is doing.

The first obvious distinction among these programmes is that the ALGOL and FORTRAN programmes are very much more con-

Table 4. FORTRAN programme to calculate the mean of ten values and decide whether or not the mean is greater than 100

```
PROGRAM MEAN
              INTEGER I
              REAL TOT,VAR, MEAN
              TOT=0
              DO 200 I=1,10
              READ(50,100)VAR
     100      FORMAT(F5.2)
              TOT = TOT + VAR
     200      CONTINUE
              MEAN=TOT/10.0
              IF(MEAN.GT.100.0)GO TO 400
              WRITE(51,300)
     300      FORMAT (1H ,18HMEAN WITHIN BOUNDS)
              GO TO 600
     400      WRITE(51,500)
     500      FORMAT(1H ,18HMEAN OUT OF BOUNDS)
     600      STOP
              END
```

Table 5. ALGOL programme to calculate the mean of ten values and decide whether or not the mean is greater than 100

```
Begin      real total, value, mean; integer count;
           total: = 0.0
           for count: =1 step 1 until 10 do
               begin input (5, '5D.2D', value); total: = total + value end;
           mean: = total/10.0;
           if mean 100.0 then output (6, "mean out of bounds")
                         else output (6, "mean within bounds")
end
```

cise than the COBOL one. The main reason for this difference in length is the extensive requirement in COBOL for defining the nature of the variables, both primary and intermediate, before any manipulation of these quantities can be carried out in the programme. In addition, all data handling within a COBOL programme has to take place by the manipulation of data records in files. These files, their nature and the particular records that they contain, together with the format designations (the "picture" statements), have to be defined before any computational procedures can be entered into. This Data Division, together with the Identification and Environment Divisions which provide descriptions of the programme and the machinery for which it is prepared, make up more than half of the COBOL programme. In a very extensive programme these may not form such a large proportion of the code, but in small computationally-based programmes such as the one being ex-

Table 6. COBOL programme to calculate the mean of ten values and decided whether or not the mean is greater than 100

```
IDENTIFICATION DIVISION
PROGRAM ID. CALCULATE-AND-TEST-MEAN.
AUTHOR. R.L. KITCHING.
INSTALLATION. GUAES.
DATE-WRITTEN. FEB 1 1978.
DATE-COMPILED. FEB 28 1978.
ENVIRONMENT DIVISION.
CONFIGURATION SECTION.
SOURCE-COMPUTER. IBM-360.
OBJECT-COMPUTER. IBM-360.
INPUT-OUTPUT SECTION.
FILE-CONTROL.
        SELECT VARIABLE-FILE ASSIGN TO CARD-READER.
        SELECT OUTPUT-FILE ASSIGN TO PRINTER.
DATA DIVISION.
FILE SECTION.
FD      VARIABLE-FILE
        LABEL RECORDS ARE OMITTED
        DATA RECORD IS VARIABLE.
01      VARIABLE     PIC IS X(5).
FD      OUTPUT-FILE
        LABEL RECORDS ARE OMITTED
        DATA RECORD IS OUTPUT-LINE.
01      OUTPUT-LINE.
        02    MESSAGE-OUT       PIC IS X(18).
WORKING-STORAGE SECTION.
77      RUNNING-TOTAL            PIC IS X(6).
77      LOOP-COUNT               PIC IS X(2).
77      YES-MESSAGE              VALUE IS MEAN WITHIN BOUNDS.
77      NO-MESSAGE               VALUE IS MEAN OUT OF BOUNDS.
77      MEAN                     PIC IS X(6).
PROCEDURE DIVISION.
INITIAL-STEPS.
        OPEN INPUT VARIABLE-FILE.
        OPEN OUTPUT OUTPUT-FILE.
        MOVE ZERO TO RUNNING-TOTAL.
        MOVE ZERO TO LOOP-COUNT.
GET-NEXT-NUMBER.
        ADD 1 TO LOOP-COUNT.
        IF LOOP-COUNT IS GREATER THAN 10 GO TO COMPUTE-MEAN.
        READ VARIABLE-FILE.
INCREMENT-TOTAL.
        ADD VARIABLE TO RUNNING-TOTAL.
        GO TO GET-NEXT-NUMBER.
COMPUTE-MEAN.
        DIVIDE 10 INTO RUNNING TOTAL GIVING MEAN ROUNDED.
MAKE-DECISION.
        IF MEAN IS GREATER THAN 100 MOVE NO-MESSAGE TO
        MESSAGE-OUT ELSE MOVE YES-MESSAGE TO MESSAGE-OUT.
WRITE-OUT-ANSWER.
        WRITE OUTPUT-LINE AFTER ADVANCING 1 LINE.
CLOSE-DOWN.
        CLOSE VARIABLE-FILE,OUTPUT-FILE.
        STOP RUN.
```

amined, they form a substantial portion of the complete coding. In comparison, the definition of the type of variables used in the FORTRAN and ALGOL programmes is in the form of declaration statements (lines 2 and 3 in the FORTRAN programme, line 1 in the ALGOL programme). In the case of the FORTRAN programme, these declaration statements can also be omitted, and conventions associated with the naming of variables can be used to provide an implicit typing of variables within the body of the programme without any separate defining statements. Apart from these differences in description of variables, the computational facility permissible within each programming language provides the most significant differences among the three considered here. Basically, the adding process involved in these computations is a looping procedure which, following an initial zeroing statement, is a single instruction in ALGOL, five lines of code in FORTRAN and seven lines of code in COBOL. In addition, ALGOL and FORTRAN are able to operate using cryptic mathematical names for variables with operational symbols, whereas customarily COBOL uses descriptive, English-like names for variables and computational instructions. There is an instruction in COBOL, the "COMPUTE" facility, which allows mathematical expressions to be written directly into the programme, but this is not a common usage in the language and in many ways is an unsatisfactory way of circumventing the basic inappropriateness of the language for mathematical operations.

Other differences between the languages, not apparent in these very simple programmes, relate to the facility and ability of the languages to handle arrays, data structures, subroutines and so on, and a summary of the most significant of these differences is provided in table 7. An examination of the three sets of code presented here, together with the differences identified in table 7, inevitably leads one to the conclusion that COBOL is inappropriate as a language for simulation modelling, a conclusion borne out by examination of the literature, and that both FORTRAN and ALGOL are admirably suited to the sorts of logic that are usually required in simulation exercises. The choice between FORTRAN and ALGOL then becomes a personal one. It is frequently stated that ALGOL is the most succinct and scientifically-adept of the two, but FORTRAN has the advantage of wider currency. Almost all of the models described in the remainder of this book will be FORTRAN-based, but this is more historical accident than any reflection on the inappropriateness of any one of a number of other languages such as ALGOL,

Table 7. Partial comparison of features of ALGOL, FORTRAN and COBOL programming languages

Feature	FORTRAN	ALGOL	COBOL
History	Scientific; US based	Scientific; UK based	Business-oriented; US based
Variable types	Integer, real, complex, double precision	Real, integer	Numeric literals only
Variable declaration	Implicit or declared	Must be declared	Must be declared
Arrays	Up to 3-dimensional restricted subscripting, static allocation	Many dimensional unrestricted subscripting, dynamic allocation	Up to 3-dimensional very restricted subscripting, static allocation – very cumbersome
Alphanumeric strings	Hollerith form, restrictive	"....." form, very restricted use	Non-numeric literals, easy to revise and move
Data structures	None but arrays	None but arrays	Files, records
Subroutines	Can be separately compiled, non-recursive	Not separately compiled can be recursive	Not physically separate at all
Intrinsic functions	Very many	Some	None
Mixed mode restrictions	Complex rules, some interconversion possible	None	None
Arithmetic	**, –, +, *,/, negation	↑, –, +, /, ÷, *	**, –, +, *, /, negation
Documentation	Comment lines	Comments or after 'END'	Largely self-documenting, "NOTE" sentences

Source: Constructed using information presented in Ralston 1971, which should be consulted for an extended and more technical account.

PASCAL or PL/1. PL/1, in fact, attempts to combine the best features of all three languages examined above and succeeds tolerably well, but it is accordingly more complex and difficult to learn and is frequently over-complex for the sorts of jobs that one wishes to do. PL/1, is restricted to IBM machines and has a limited availability outside the United States.

4.2.3 Structured Programming

The most common and generally useful way to overcome the problems of writing and presenting complex programmes in simulation modelling is by the adoption of an approach which has become known as "structured" or "top-down" programming. This is based on the idea that a continuous uninterrupted block of code containing the whole of a programme is hard to read and even harder to write, given the constant modifications required as part of the process of programme construction. To facilitate these activities and to introduce some logical pattern into the whole programming process, the programme can be structured. The sorts of structures used are to some extent determined by the programming language adopted but, in FORTRAN and ALGOL, would normally take the form of subroutines. These are small programmes, each carrying out a short and discrete sequence of operations on the "call" of the main or connecting programme. The rules associated with subroutines vary from language to language, but FORTRAN has a particularly useful activity in this regard as each subroutine can be compiled and debugged separately, that is, the writing of each subroutine is to some extent independent of the writing of the rest of the programme. This introduces a modularity into the finished programme which is easy to understand and straightforward to modify.

Given this facility of the subroutine structure, it has been suggested that those writing complex programmes — such as the ones usually produced by systems ecologists — would be well advised to systematize their approach to the job in a hierarchical fashion, referred to as the top-down approach. Although this notion has been current for some years now there are few accounts of the ideas involved. The short introduction given here is based on van Tassell (1974).

The approach includes the following steps:

1. The set of logical processes that it is intended to build into the programme are written down in their order of operation. This must be organized into a series of modules, each including only a very short sequence of steps. If the intended programme contains branching processes then the list of logical steps can be hierarchically structured. The smallest unit of this whole process can be equated with the subroutine or related structure of the programming language itself.
2. The data to be used by the model and the pathway of particular data items through the logic of the intended programme must be identified. This should be accompanied by hand calculations of simple examples of these procedures. This data should be subdivided in such a way that

certain portions of the data base are identifiable as appropriate test data for each of the programming modules recognized earlier.

3. Initial programming is frequently done in "skeleton" style. This means that, instead of each module being written initially in its full complexity, it is represented by a simplified or even dummy segment of code. This enables the whole programme to be constructed quickly, and the ramifications of its structures in terms of identifying critical modules, optimal module sizes, their complexity and adequacy can be explored.
4. Final programming can be done in a stepwise manner, module by module. This is particularly useful in certain sorts of ecological modelling where we have some idea of the overall structure and connectedness of our system, but where our degree of detailed knowledge may vary from area to area within the system. Having constructed a skeleton model describing the whole system, we can fill in the modules describing those areas about which we have extensive knowledge while leaving other segments until a later time. This means we begin to think about the place of the processes we model in detail in the wider context of the higher levels of resolution of systems and, at the same time, identify areas of particular ignorance.

The wider adoption of structured programming techniques would not only produce more logical programmes and ease the programmer's burden but would also lead to a uniformity of approach which would make presentation and communication of programmes much simpler. Apart from anything else, the procedures described and the modularity introduced into the programming procedures force the adequate documentation of the process — a vital but often neglected part of the programming process and the subject of the next section.

4.2.4 Documentation

I have already stressed the importance of producing programmes written in standard versions of higher level languages to facilitate transferability of these programmes between installations. Another frequent and major shortcoming in many programmes is their lack of internal and external documentation. Anyone who has picked up a programme written by someone else or even by themselves at an earlier time will appreciate the difficulties in trying to pick one's way through a mass of code with its idiosyncratic variable names, structure and style. On occasion this has even led to decisions to rewrite the programmes rather than continue struggling to understand one that already exists. In addition, there is a growing trend for the use of pre-programmed packages for particular pieces of statistical or other analyses and for the construction of programmes which draw upon sets of these prepared modules. In all normal applications, including the preparation of modules for later use, therefore, it is vital that the programmer develop good documentation skills and apply them assiduously and without exception in all his activities. Such documentation can be divided into two major categories, internal and external.

The internal documentation of a programme is used here to mean those comments which are permissible within the body of the code itself and which appear on any printout of the programme by the computer. Different higher level languages provide different facilities in this regard. The two common algebraic languages, FORTRAN and ALGOL, are particularly in need of adequate commenting whereas COBOL, because of its English-like nature, is to a large extent self-documenting, assuming careful choice of variable names has been made. FORTRAN permits the insertion of comment cards distinguished from others by having a capital "C" in the first column of each line, these may be inserted anywhere and in any number throughout a programme. In fact, the compiler will ignore them in passing on computational instructions to the computer but they will be included in any printout of the programme listing. ALGOL has a similar facility and comments may be inserted after the "begin' instruction or throughout the programme, each note being preceded by the word "comment". In addition ALGOL permits any number of comments after the "end" statement. COBOL, although largely self-documenting, also provides the facility for the insertion of "note" sentences throughout the programme code, each of which is preceded by the word "note". COBOL also has a "remarks" section in the introductory identification division of every programme, and this can be used to insert a paragraph of descriptive text referring to the programme. All of the comment facilities available in whatever language are useful for a number of reasons. They permit the programme to be introduced and described early in the code, they enable a list of variables to be printed out, each accompanied by an explanatory phrase, and they can be used to identify specific blocks of code within the computational flow. Their use, however, is limited, and external documentation is usually essential as an adjunct to thesc internal devices.

External documentation can take a number of forms, some or all of which should accompany each completed programme. These can be dealt with under four headings:
- Flowcharts
- Narrative descriptions
- User-orientated instructions
- Operating instructions

Flowcharts

Flowcharts are a diagrammatic way of describing a computer programme and are the most useful single piece of documentation that can accompany any programme. They can be drawn at a number of levels, variously referred to as macro-flowcharts, micro-flowcharts, high level flowcharts and so on. Usually at least two levels of flowcharts are required to describe a programme adequately. These use a common set of symbols and figure 14 shows the standard versions of these symbols. Flows of computations and information within the programme are represented by single lines and convention dictates that unless absolutely necessary these should be directed down the page and from left to right. These flows connect a series of boxes within which computing instructions are written. The nature of these instructions dictates the shape of the box concerned and the various shapes are illustrated in the figure. The basic design of the programme, both before it is written and as part of its later documentation, is often best represented as a macro-flowchart. This contains the outline of the logic involved in the construction of the programme without going into the detail required before the programme itself is written. This identifies starting points and finishing points, major blocks of computation within the programme and major branches and the decisions that accompany these branching processes.

From the point of view of documentation and understanding a programme, the user will find a micro-flowchart most useful. This level of flowchart represents in some detail the logic of the programme itself. The programme that we have already examined in tables 4, 5 and 6 can be represented as a micro-flowchart in the fashion shown in figure 15. The start of the programme is indicated by a box with rounded ends from which the computational flow passes to a preparation box before we enter the main computational loop. This box is polygonal and in this case con-

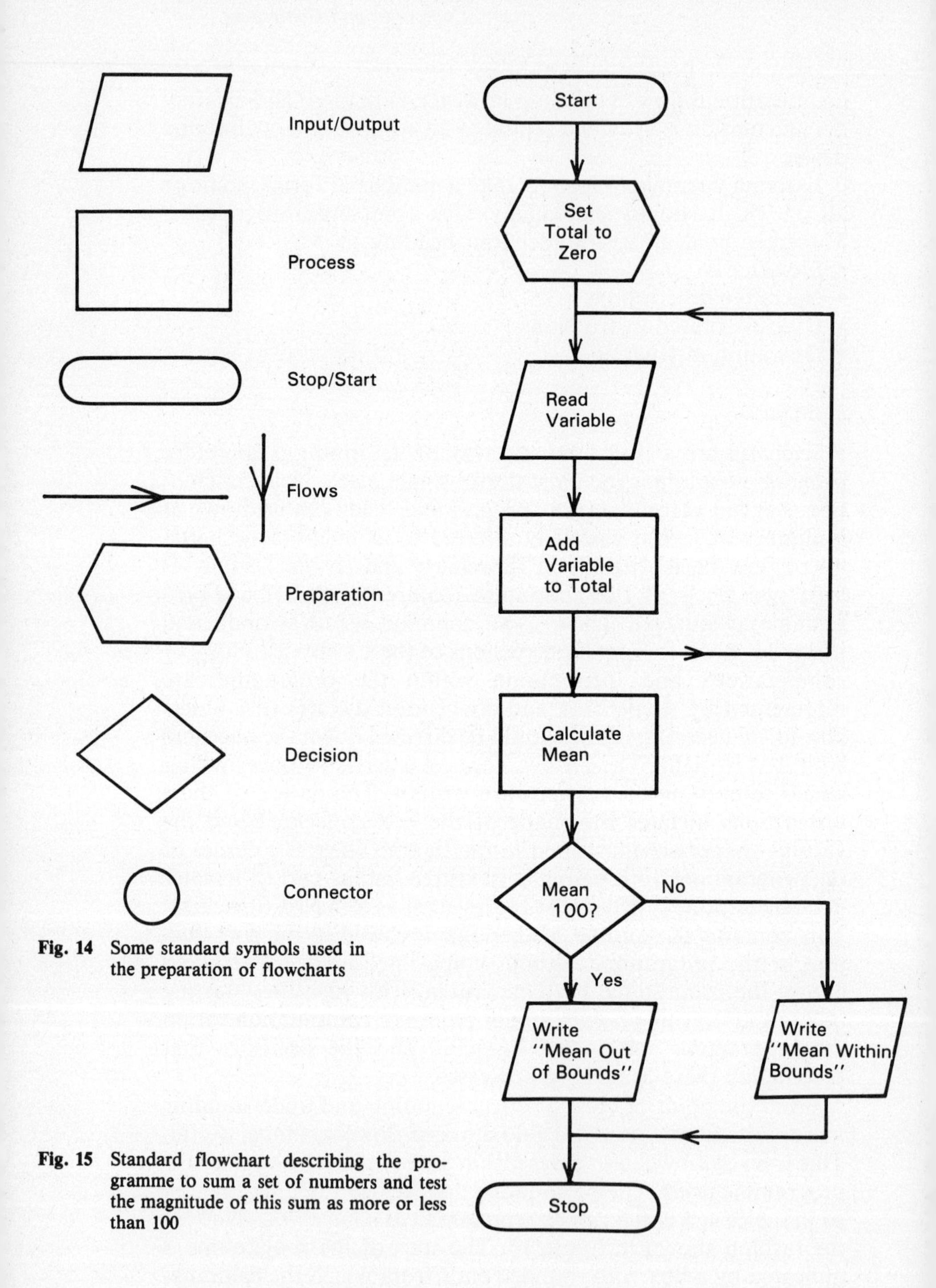

Fig. 14 Some standard symbols used in the preparation of flowcharts

Fig. 15 Standard flowchart describing the programme to sum a set of numbers and test the magnitude of this sum as more or less than 100

tains the instructions relating to zeroing of the running total. We then enter the main loop in which the running addition is carried out and each pass through the loop requires two operations: the reading in of a variable which is represented by a box with parallel sloping sides, and the addition of this variable to the total, which is represented in a rectangular process box. Leaving the adding loop, we must calculate the mean, and this instruction is also contained in a process box. The decision with respect to the calculated mean is indicated in the flowchart by the means of a diamond-shaped decision box. Decision boxes in general must have at least two outlets from them, representing the simple yes/no decision, although they can be more complex and have three or many outlets. In the example given, a simple yes/no decision is involved and, depending upon whether or not the calculated mean is greater than one hundred, the flow passes to one of the two "write" statements instructing the programme to produce the appropriate message. "Write" statements occur in flowcharts as an input/output box of the same kind as was used for the "read" instruction, having sloping parallel sides. This is the last activity within this particular programme, and the computational flow passes from both "write" boxes to a "stop" instruction contained in a box comparable with that used at the start of the programme. This very simple programme has shown all the common features of flowchart preparation with the exception of one, the connector symbol. This is represented by a number or a letter in a circle and is used when the flowchart being produced is complex and many-stranded. If, for instance, we wish to indicate that a particular decision in our flow of computations leads to a set of code which is flowcharted separately, then we might use an arrow leading to a connector box labelled A or B or some other symbol, and indicate on a separate flowchart the entry from such a point by the use of another connector symbol containing the same label. Examples of the more complex structuring will be found in chapters 6, 7 and 8, and one or more flowcharts are presented for most of the models discussed.

Narrative description

The flowchart is a pictorial representation of what a programme does. It is often desirable, however, to have a written description of the aims and functioning of a particular programme, and this is a useful part of the documentation process. A narrative description is easier to read than the flowchart, but on the other hand can be more ambiguous. The principles of the preparation of such a description are the same as those pertaining to all scientific writing. They should be clear, in simple terms and should be suitably structured under headings and sub-headings. The narrative description should be used hand in hand with the flowcharts prepared to describe the same programme, the two being different aspects of the same descriptive process.

User-oriented instructions

This part of the documentation should be in the form of instructions as to exactly how the programme concerned is used, what input is required, what format this input must take, explanation of the various intermediate output messages that may occur, the sorts of output that may be expected as the end product of the programme and, lastly, the nature of the calculations involved. Essentially, what is involved is a set of instructions as to precisely how the new user can operate the programme while knowing nothing about that specific set of procedures before he starts.

Operating Instructions

Various input/output and storage devices are available in computer installations, and particular programmes call upon different sets of devices. It is necessary to know before a programme is run exactly what devices are required, and operating documentation should include details of the input, output and storage devices required. This documentation should also indicate any temporary storage requirements in the form of disc or drum space and the fate of these storage modules after the programme has been executed.

The above account has dealt briefly with the main forms of documentation which most programmes require in order to be useful to users other than the original programmer. Other documentation in the form of references, key descriptions of background information, examples of the output and input, instructions about alternative forms of output that can be obtained and possible developments that a future user may wish to build into the model can all be added to this body of documentation under certain circumstances. Ultimately, the form that documentation takes with respect to a particular programme is an idiosyncratic decision of the programmer. However, a particular body of basic knowledge is essential if the programme is to be anything other than an entirely personal tool.

5

Some Mathematics

As mentioned in chapter 3, there are a great many areas of mathematics that have been applied profitably to the description and modelling of ecological systems. To attempt a complete treatment of even a small number of these would be inappropriate in a work written by a biologist for biologists, although to omit entirely any treatment of mathematics would be to deny the essential mathematical nature of systems analysis. Introductory accounts are therefore given of four areas of technique, some understanding of which is vital to the appreciation of the accounts of models which are presented in later chapters. These four areas all relate to the representation and summation of the transitions of a system or parts thereof from one state to another. The first two, **differential and difference equations of change** and some aspects of **linear algebra**, are directed at actually summarizing the numerical changes which occur in one or more state variables. The third topic, on **stability**, is concerned more with the description of patterns or change in systems through time and the persistence or demise of particular components over any period. Lastly, a brief account of notions of **probability** is given as assistance in thinking about stochastic rather than deterministic representation of dynamical changes within systems.

These are minimal accounts designed to help the largely non-numerate overcome one more barrier in the approach to complex ecological models; they do not purport to do more than equip the reader with the most basic concepts and terminology. Throughout, references to more complete, essentially mathematical, treatments are provided for the interested student. I have deliberately omitted any account of one most important tool for the systems analyst, namely statistics. This is done not only because of considerations of balance in an essentially biological work, but also because a wide range of works are available, from "cookbooks" to near complete theoretical treatments on statistics; in addition, most biologists will have at least some acquaintance with the topic. Jeffers (1978) provides a useful account connecting the more conventional use of statistics in experimental design and analysis, with its use in systems ecology.

5.1 Equations of Change

If we measure the values of any two variables observed in the natural world at the same time, then it is possible to draw a graph of one such variable against the other. If there is a relationship, direct or indirect, between the two variables, the points obtained on a simple two-dimensional graph of this sort will describe some pattern — a straight line, a rising curve, an S-shaped curve and so on. The functional relationship between the two variables can be expressed by obtaining the equation for this line by analytical or statistical techniques. The systems ecologist combines many such relationships when he builds his models of particular ecological systems. However, as his primary interest is in the dynamics of the system through time, he is often more interested in the rate equation representing the slope of the line relating a particular variable to time. Equations relating the rate of change of variables (their **differential** or **first derivatives**) to the values of the variables themselves are called **differential equations**, and are to be distinguished from the integral equations relating the levels of particular variables to time itself. This terminology and the form and manipulation of the equations are based on the infinitesimal calculus.

Figure 16 shows a series of examples of differential equations of increasing complexity based on different forms of population growth. The graphs on the left-hand side of the figure show the integral form of the relationship, relating the total number of individuals in a population (N) to time (t), and those on the right-hand side show the differential form, relating the rate of change in population numbers (dN/dt) to N. Graphs a and b illustrate the **linear** growth situation, where the relationship between N and t can be represented by a straight line:

$$N = rt$$

where r, the slope of the line, is the rate of increase in numbers per unit time. The differential of this:

$$\frac{dN}{dt} = r$$

is a rather uninteresting form as, of course, the rate of increase in numbers is constant for the linear case. This example

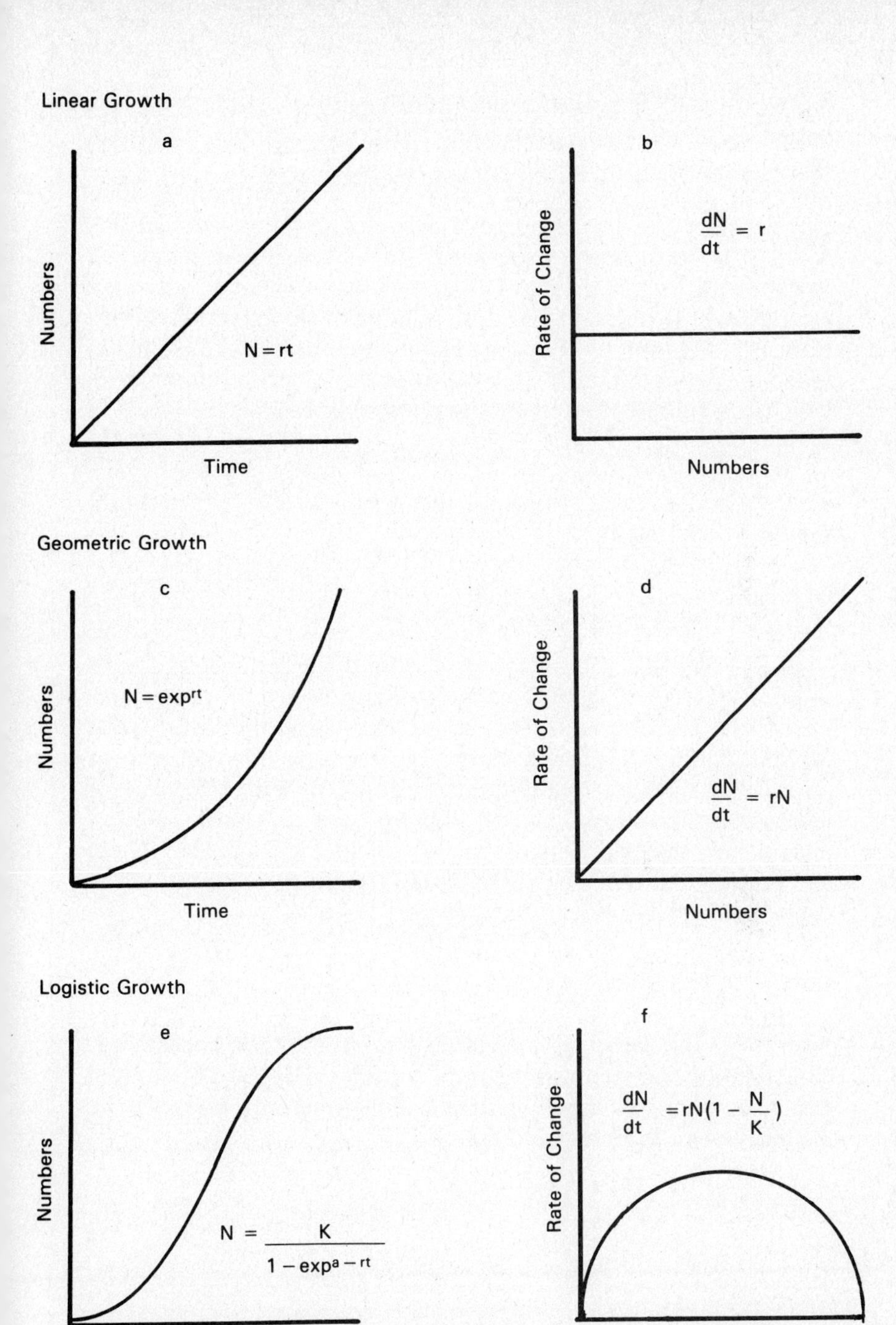

Fig. 16 Some examples of integral curves (on the left) for particular forms of population growth, with the corresponding graphs of the differential curves involved (on the right)

represents an unrealistic situation where a population experiences no mortality, does not reproduce and yet receives a steady stream of immigrants from its surroundings. The second example, shown in graphs c and d, is a more realistic one, representative of a wide range of pest or immigrant populations which, to start with, experience an effectively unrestrictive environment (see Hardman 1976a and chapter 6 for an extended treatment of this sort of situation in pests of stored products). In this case the graph of N against t shows a **non-linear** relationship indicative of **geometric** (or **exponential)** population growth, which can be represented by the integral equation:

$$N = e^{rt}$$

where e is the base of natural logarithms (2.7182). The more familiar differential form is linear:

$$\frac{dN}{dt} = rt.$$

This is an ordinary, linear, first order differential equation of the first degree.* Graphs e and f in figure 16 are the most complex to be considered at this stage, and show the well-known **logistic** (sigmoid) growth form representing the pattern of population increase in a limited environment where the rate of increase is non-linear, diminishing to zero as the population approaches the "carrying capacity" of its environment. There are many forms of the differential equation involved, but a useful and easily understood one is:

$$\frac{dN}{dt} = rN\left(1 - \frac{N}{K}\right)$$

where K is the carrying capacity experienced by the population. This is only one step more complex than the linear differential equation, being a simple quadratic. This is perhaps clearer if we multiply out the right-hand side:

$$\frac{dN}{dt} = rN - \frac{r}{K}N^2$$

---

*"Ordinary" because only one variable other than time is present on the left-hand side — compare with "partial"; "first order" because the equation contains only a first derivative (dN/dt); "first degree" because there are no powers of derivatives involved other than 1. These distinctions can be explored in any standard text on differential equations such as Goult *et al.* (1973) or Boyce and DiPrima (1965), but will not be dealt with further in this work.

(Compare this with the simple algebraic quadratic, $y = a + bx + cx^2$ where $a = 0$, $b = r$ and $c = -r/K$.)

The integral form of the logistic is complex and its derivation will not be given here. Graph e shows one version of this equation where a is the constant of integration.

Differential equations are continuous in nature, consistent with their link with the Newtonian calculus. However, many ecologists and other workers have preferred to use the discrete analogues of differential equations, namely **difference equations**. Once again it is only necessary to explain the essential nature of these tools here and this is best done by comparison with the differential equations already explored. The basic mathematical difference is that whereas the differential equations denote changes in time over the infinitesimal interval:

$$t \rightarrow t + \delta t$$

difference equations represent discrete changes over the finite interval:

$$t \rightarrow t + \Delta t.$$

Table 8. Differential and difference equation forms for population growth (see also table 9)

Type	Differential form	Difference form
Linear growth	$\frac{dN}{dt} = r$	$N_{t+\Delta t} = N_t + r\Delta t$
Geometric growth	$\frac{dN}{dt} = rN$	$N_{t+\Delta t} = N_t + rN_t\Delta t$
Logistic growth	$\frac{dN}{dt} = rN(1 - \frac{N}{K})$	$N_{t+\Delta t} = N_t + rN_t\Delta t(1 - \frac{N_t}{K})$

Note: See also table 9.

Very simply this means that a summation procedure replaces the integration used in manipulating differential equations. Table 8 compares the examples of equations of population growth already covered with the analogous difference forms. In population studies there is a body of opinion which favours the use of difference equations for situations where generations overlap and differential equations for the effectively continuous situation of overlapping generations. An introduction to the mathematics of difference equations is given by Goldberg (1961), and May (1975, 1981a, 1981b) provides extensive comment on their applications in population ecology.

The nature of simple differential and difference equations should now be clear and we can turn to the ways in which they are applied by systems ecologists. Before doing this however, it is necessary to explain the nature of the **solution** of such equations. The many techniques developed by mathematicians for solving differential equations are directed to the finding of the integral equations associated with particular forms of equations, that is, the algebraic relationship between y and x which satisfied the equation. There is a great variety of analytical techniques involved, introductions to which can be found in texts such as Leighton (1963), Boyce and DiPrima (1965) and many others. For difference equations there is a smaller but growing body of technique directed to finding a function which satisfies the particular relationship involved in the difference equations for all intervals over which the equation is defined (Goldberg 1961). However, an equation of either kind does not need to be very complex to defy even the most sophisticated techniques of analytical solution. For this reason, a large body of **numerical** methods have been developed which use the computational power of computers to find solutions to differential and difference equations using "cut and fit" techniques. For differential equations this is achieved by converting to a difference form using various methods, the simplest of which is laid out in table 9. To "solve" the equation numerically we calculate a series of values of N_t through time and plot them. There is, however, an infinite series of such solution curves (isoclines), each differing

Table 9. The conversion of differential to difference equations

Consider the logistic equation

$$\frac{dN}{dt} = rN(1 - \frac{N}{K})$$

Let us approximate the differential for the time interval $t \rightarrow t + \Delta t$ by the straight-line average

$$\frac{N_{t+\Delta t} - N_t}{\Delta t}$$

Note that the accuracy of the approximation depends of the size of Δt.
Then we can write the equation

$$\frac{N_{t+\Delta t} - N_t}{\Delta t} = rN_t(1 - \frac{N_t}{K})$$

and rearrange this to the usual difference form

$$N_{t+\Delta t} = N_t + rN_t\Delta t(1 - \frac{Nt}{K})$$

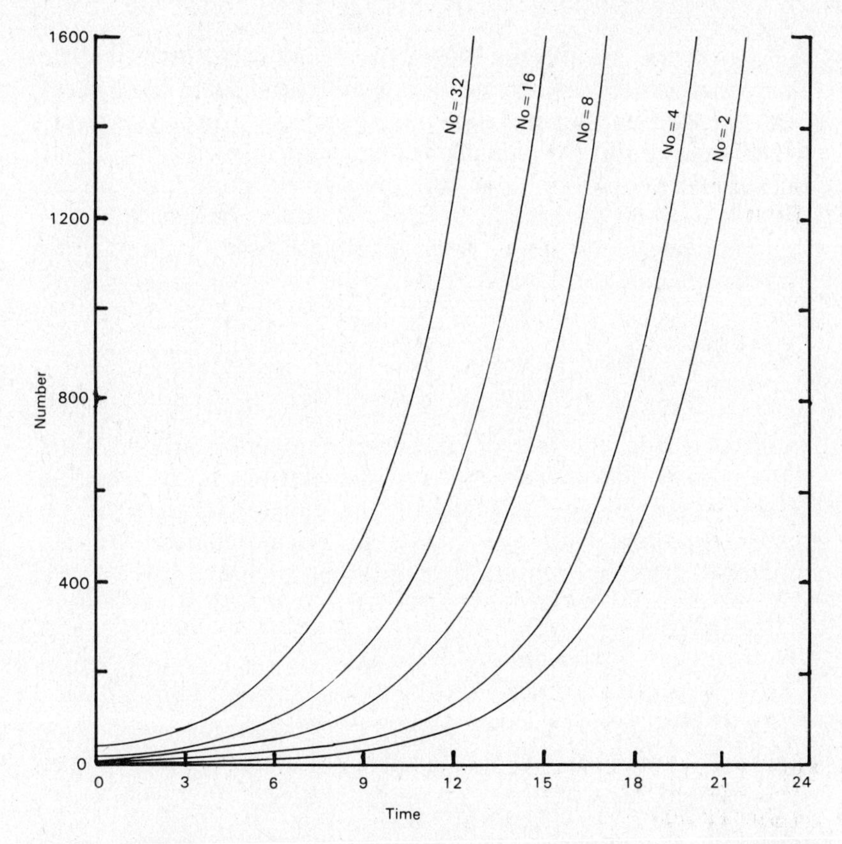

Fig. 17 Solution curves for the geometric growth equation $N_{t+\triangle t} = N_t + rN_t\triangle t$ with r set to 0.5 and $\triangle t$ to 3. Curves for initial values N_o of 2, 4, 8, 16 and 32 are shown

in the starting point that we used in our evaluation (that is, N_t at $t = 0$). To relate a particular solution curve to a particular equation we must know the starting conditions involved. Figure 17 shows a few of the infinite set of solution curves for the geometric growth equation discussed above. Any one of these represents a solution to the equation concerned, but only one is appropriate for the particular population under study. Thus, if we have reasonable grounds for expecting our population to grow geometrically and if we start with two individuals, then we might expect that it will follow curve one in the figure; if we start with four individuals, curve two; eight, curve three and so on — the "correct" solution curve is defined by the starting value.

We have spoken so far of single equations as the system of interest. The systems ecologist is interested, of course, in sets of such equations containing the relationships which describe the changes that occur in each of the state variables in the system being worked on. For instance, the population ecologist is likely to be interested in the changes which occur in each age class of his subject population, not just the overall abundance of the species, be it egg or imago, juvenile or adult. We can illustrate this mathematically for differential equations by putting them into the general form, say:

$$\frac{dN}{dt} = f(N)$$

In other words, the rate of change in numbers is related functionally to N although the form of the function is not specified at this stage, being indicated by the general term, f. If we recognize, now, four age classes in the population we can replace this single equation by the system of four:

$$\frac{dN_1}{dt} = f(N_1, N_2, N_3, N_4)$$

$$\frac{dN_2}{dt} = g(N_1, N_2, N_3, N_4)$$

$$\frac{dN_3}{dt} = p(N_1, N_2, N_3, N_4)$$

$$\frac{dN_4}{dt} = q(N_1, N_2, N_3, N_4)$$

This says that the rates of change of the four age classes, N_1, N_2, N_3 and N_4 are related to the numbers of animals in these four classes by the functions, f, g, p and q. The functions concerned can be of varying complexity, linear or non-linear, and to "solve" the system of equations we must use numerical methods which enable us to calculate the changes in values of the state variables, N_1, N_2, N_3 and N_4 through time. The methods involved operate on the same principles as have been outlined already and again initial values are critical. With even as few as four equations involved the manipulations are cumbersome and time consuming. Fortunately some of the labour can be alleviated using the terminology and techniques of linear algebra, which is discussed in the next section.

5.2 Vectors, Matrices and their Manipulation

We have already encountered the notion of linked systems of equations; let us look more closely at a pair of very simple algebraic equations:

$$4x + 3y + 5z = 11$$
$$2x + 4y + z = 5$$

These equations are **simultaneous** in the variables x, y and z and their left-hand sides can be summarized by the rectangular array of coefficients:

$$\begin{matrix} 4 & 3 & 5 \\ 2 & 4 & 1 \end{matrix}$$

and the right-hand side by the simpler array:

$$\begin{matrix} 11 \\ 5 \end{matrix}$$

Such arrays are referred to as **matrices** and the one dimensional case as a **vector**. In algebraic notation, a matrix is usually represented by a bold capitalized letter; for instance, we might call the coefficient matrix of our pair of equations **C**. A single dimensioned matrix, a vector, is written using the notation $\underline{s}$, $\vec{s}$ or $\utilde{s}$ or, simply **s**, where the single letter represents the vector. In this account we shall use the last of these notations. In longhand the sets of **elements** which make up a matrix or vector are enclosed in large brackets:

$$\begin{bmatrix} 4 & 3 & 5 \\ 2 & 4 & 1 \end{bmatrix}$$

Each matrix is made up of a number of rows and columns and is usually referred to as an (m x n) matrix, where the first number, m, is the number of rows and the second, n, the number of columns. When m or n equals one we have either a **row** or a **column vector**. The case of both m and n being one is degenerate and is just a single number, a **scalar** as opposed to a matrix. Any element of a matrix can be denoted by its row and column number and is given the same letter as the matrix although in normal face, lower case type. Thus any element of the matrix **C** is referred to as c_{ij} where i and j are row and column positions respectively, of the particular element. In general then we may write the (m x n) matrix, **C**, as shown.

$$\mathbf{C} = \begin{bmatrix} c_{11} & c_{12} & c_{13} & \dots & c_{1n} \\ c_{21} & c_{22} & c_{23} & \dots & c_{2n} \\ \cdot & \cdot & \cdot & & \cdot \\ \cdot & \cdot & \cdot & & \cdot \\ \cdot & \cdot & \cdot & & \cdot \\ \cdot & \cdot & \cdot & & \cdot \\ c_{m1} & c_{m2} & c_{m3} & \dots & c_{mn} \end{bmatrix}$$

One matrix equals another only when m and n are the same for both, and every element is equal to the corresponding one of the other matrix.

These notions can be put into an ecological context by reconsidering the age-structured model comprised of four differential equations given at the end of 5.1. The left-hand sides of the equation can now be rewritten as a column vector:

$$\begin{bmatrix} dN_1/dt \\ dN_2/dt \\ dN_3/dt \\ dN_4/dt \end{bmatrix} = \mathbf{d}$$

Table 10. An age-structured population model and its matrix representation

Let a population be comprised of four age classes whose numbers at any point in time are N_1, N_2, N_3 and N_4. The changes in these numbers through time can be expressed generally as the set of equations*:

$$\begin{aligned} N_1(t+\Delta t) &= N_1(t) + \Delta t.f\,[N_1(t), N_2(t), N_3(t), N_4(t)] \\ N_2(t+\Delta t) &= N_2(t) + \Delta t.g\,[N_1(t), N_2(t), N_3(t), N_4(t)] \\ N_3(t+\Delta t) &= N_3(t) + \Delta t.p\,[N_1(t), N_2(t), N_3(t), N_4(t)] \\ N_4(t+\Delta t) &= N_4(t) + \Delta t.q\,[N_1(t), N_2(t), N_3(t), N_4(t)] \end{aligned}$$

Now if each age class gives rise to a number of offspring of age N_1 in a particular time interval (and the birth rate per individual is represented by b_1, b_2, b_3 and b_4) and the proportion surviving from one age class to the next is S_{12}, S_{23}, S_{34} respectively, we can rewrite these equations as:

$$\begin{aligned} N_1(t+\Delta t) &= b_1.\,N_1(t) + b_2.\,N_2(t) + b_3.\,N_3(t) + b_4.\,N_4(t) \\ N_2(t+\Delta t) &= S_{12}.\,N_1(t) + 0.\,N_2(t) + 0.\,N_3(t) + 0.\,N_4(t) \\ N_3(t+\Delta t) &= 0.\,N_1(t) + S_{23}.\,N_2(t) + 0.\,N_3(t) + 0.\,N_4(t) \\ N_4(t+\Delta t) &= 0.\,N_1(t) + 0.\,N_2(t) + S_{34}.\,N_3(t) + 0.\,N_4(t) \end{aligned}$$

Note that the Δt is such that an animal must either have died or aged into the next age class (hence the removal of the $N_i(t)$ terms from the explicit forms of the equations). These equations can be rewritten in matrix form as:

$$\begin{bmatrix} N_1(t+\Delta t) \\ N_2(t+\Delta t) \\ N_3(t+\Delta t) \\ N_4(t+\Delta t) \end{bmatrix} = \begin{bmatrix} b_1 & b_2 & b_3 & b_4 \\ S_{12} & 0 & 0 & 0 \\ 0 & S_{23} & 0 & 0 \\ 0 & 0 & S_{34} & 0 \end{bmatrix} \begin{bmatrix} N_1(t) \\ N_2(t) \\ N_3(t) \\ N_4(t) \end{bmatrix}$$

* *Observe that when some subscript is needed in addition to the time one, it is convenient to use the designation N(t) rather than N_t in a difference equation*

the coefficients of the functions as a 4 x 4 matrix, **T** say, and the variables involved in the right-hand side as another vector:

$$\begin{bmatrix} N_1 \\ N_2 \\ N_3 \\ N_4 \end{bmatrix} = \mathbf{n}$$

We can then rewrite the whole system

$$\mathbf{d} = \mathbf{T}\,\mathbf{n}$$

Implicit in this notion is the idea that matrices and vectors can be manipulated arithmetically and we shall return to the methods involved shortly. Table 10 provides an example of a set of equations of this type and their matrix form. For convenience the equations are expressed in difference form. Note the great economy that is achieved using matrix notation.

Matrix arithmetic

Two matrices can be added or subtracted only if they are of the same **order**; that is, if each has the same number of rows and columns as the other. Thus the matrices

$$\begin{bmatrix} a & b \\ c & d \end{bmatrix} \quad \text{and} \quad \begin{bmatrix} w & x \\ y & z \end{bmatrix}$$

can be added or subtracted whereas the matrices

$$\begin{bmatrix} a & b \\ c & d \end{bmatrix} \quad \text{and} \quad \begin{bmatrix} u & v \\ w & x \\ y & z \end{bmatrix}$$

are incompatible for these operations. These processes are achieved by adding or subtracting corresponding elements in the two compatible matrices. Table 11 gives some numerical and algebraic examples of matrix addition and subtraction. Matrix addition and subtraction are both commutative and associative. That is to say if we have three matrices **A, B,** and **C** then:

$$\mathbf{A}+\mathbf{B}+\mathbf{C} = (\mathbf{A}+\mathbf{B})+\mathbf{C}$$

In addition where **0** is the zero matrix, all of whose elements equal zero:

$$\mathbf{A}+(-\mathbf{A}) = \mathbf{0} \quad \text{and} \quad \mathbf{A}+\mathbf{0} = \mathbf{A}$$

Table 11. Examples of matrix addition and subtraction

(a) $\begin{bmatrix} 3 & 4 \\ 5 & 6 \end{bmatrix} + \begin{bmatrix} 1 & 3 \\ -4 & 6 \end{bmatrix} = \begin{bmatrix} 4 & 7 \\ 1 & 12 \end{bmatrix}$

(b) $\begin{bmatrix} 3 & 4 \\ 5 & 6 \end{bmatrix} - \begin{bmatrix} 1 & 3 \\ -4 & 6 \end{bmatrix} = \begin{bmatrix} 2 & 1 \\ 9 & 0 \end{bmatrix}$

(c) $\begin{bmatrix} 1 \\ 4 \\ 6 \end{bmatrix} + \begin{bmatrix} 10 \\ 25 \\ -10 \end{bmatrix} = \begin{bmatrix} 11 \\ 29 \\ -4 \end{bmatrix}$

(d) $\begin{bmatrix} a & b \\ c & d \end{bmatrix} + \begin{bmatrix} w & x \\ y & z \end{bmatrix} = \begin{bmatrix} a+w & b+x \\ c+y & d+z \end{bmatrix}$

Two matrices can be multiplied together only if the number of columns in one is the same as the number of rows in the other. If we multiply an (n x k) matrix by a (k x m) matrix the product will be an (n x m) matrix, each element of which is the sum of the products of the elements in the ith **row** of the first with the corresponding elements in the jth **column** of the second.

In general terms, if we have two matrices **A** and **B** of order (3 x 3) and (3 x 2) with elements a_{ij} and b_{ij} respectively, the elements (c_{ij}) of the product matrix **C** will each be defined as follows:

$$c_{ij} = a_{i1}.b_{1j} + a_{i2}.b_{2j} + a_{i3}.b_{3j}$$

A numerical example of the multiplication of two matrices of this order is given in table 12. This process is of particular importance to the systems modeller, as it is through matrix multiplication that he can predict the dynamics of systems behaviour, using the basic transformation equation referred to in chapter 2. This is best explained be returning to our example

Table 12. An example of matrix multiplication

Multiply **A** by **B** where

$$\mathbf{A} = \begin{bmatrix} 4 & 2 & 4 \\ -1 & 0 & 2 \\ 3 & 6 & 5 \end{bmatrix} \quad \text{and} \quad \mathbf{B} = \begin{bmatrix} 1 & 0 \\ 2 & 6 \\ 3 & 5 \end{bmatrix}$$

These are of order (3 x 3) and (3 x 2) so we expect a product matrix, **C**, of order (3 x 2). Thus

$$\mathbf{A.B} = \begin{bmatrix} (4x1)+(2x2)+(4x3) \\ (-1x1)+(0x2)+(2x3) \\ (3x1)+(6x2)+(5x3) \end{bmatrix} \bullet \begin{bmatrix} (4x0)+(2x6)+(4x5) \\ (-1x0)+(0x6)+(2x5) \\ (3x0)+(6x6)+(5x5) \end{bmatrix}$$

$$= \begin{bmatrix} 4+4+12 & 0+12+20 \\ -1+0+6 & 0+0+10 \\ 3+12+15 & 0+36+25 \end{bmatrix}$$

$$= \begin{bmatrix} 20 & 32 \\ 5 & 10 \\ 30 & 61 \end{bmatrix} = \mathbf{C}$$

Table 13. An example of matrix representation of the ageing of a population

Let the birth rate per individual per time period for each of the four age classes in a population be 0, 0.5, 2.0, 1.0, and the proportions surviving from one class to the next be, respectively, 0.8, 0.6 and 0.3. Then we can write the model described in table 10 as:

$$\begin{bmatrix} N_1(t+\Delta t) \\ N_2(t+\Delta t) \\ N_3(t+\Delta t) \\ N_4(t+\Delta t) \end{bmatrix} = \begin{bmatrix} 0 & 0.5 & 2.0 & 1.0 \\ 0.8 & 0 & 0 & 0 \\ 0 & 0.6 & 0 & 0 \\ 0 & 0 & 0.3 & 0 \end{bmatrix} \begin{bmatrix} N_1(t) \\ N_2(t) \\ N_3(t) \\ N_4(t) \end{bmatrix}$$

If our starting vector $\mathbf{n_0}$ is, say, (10 10 10 10), the first transition will be calculated as follows:

$$\begin{bmatrix} 0 & 0.5 & 2.0 & 1.0 \\ 0.8 & 0 & 0 & 0 \\ 0 & 0.6 & 0 & 0 \\ 0 & 0 & 0.3 & 0 \end{bmatrix} \begin{bmatrix} 10 \\ 10 \\ 10 \\ 10 \end{bmatrix} = \begin{bmatrix} (10x0)+(10x0.5)+(10x2.0)+(10x1.0) \\ 0.8x10 \\ 0.6x10 \\ 0.3x10 \end{bmatrix} = \begin{bmatrix} 35 \\ 8 \\ 6 \\ 3 \end{bmatrix}$$

In the same fashion we can multiply the new vector by the transition matrix and proceed to obtain the sequence of vectors that represents the ageing dynamics of the population:

$$\begin{bmatrix} 10 \\ 10 \\ 10 \\ 10 \end{bmatrix} \rightarrow \begin{bmatrix} 35 \\ 8 \\ 6 \\ 3 \end{bmatrix} \rightarrow \begin{bmatrix} 19 \\ 28 \\ 5 \\ 1 \end{bmatrix} \rightarrow \begin{bmatrix} 25 \\ 15 \\ 17 \\ 1.5 \end{bmatrix} \rightarrow \begin{bmatrix} 43 \\ 20 \\ 10 \\ 0.5 \end{bmatrix}$$

and so on.

of an age-structured population described in table 10. If actual values are assigned to the b's and s's we can compute the change in age structure from t to t + Δ t. This example is enumerated in table 13.

Table 14. The multiplication of a matrix by a scalar

For any matrix **A** of order (m x n) with elements a_{ij}, multiplication by any scalar λ is achieved as follows:

$$\lambda \cdot \begin{bmatrix} a_{11} & a_{12} & a_{13} & \ldots a_{1n} \\ a_{21} & a_{22} & a_{23} & \ldots a_{2n} \\ \vdots & \vdots & \vdots & \vdots \\ a_{m1} & a_{m2} & a_{m3} & \ldots a_{mn} \end{bmatrix}$$

$$= \begin{bmatrix} \lambda a_{11} & \lambda a_{12} & \lambda a_{13} & \ldots \lambda a_{1n} \\ \lambda a_{21} & \lambda a_{22} & \lambda a_{23} & \ldots \lambda a_{2n} \\ \vdots & \vdots & \vdots & \vdots \\ \lambda a_{m1} & \lambda a_{m2} & \lambda a_{m3} & \ldots \lambda a_{mn} \end{bmatrix}$$

Example

To multiply the matrix $\begin{bmatrix} 6 & -1 & 3 \\ 4 & 0 & 2 \\ 1 & 8 & 12 \end{bmatrix}$ by the scalar 7

$$7. \begin{bmatrix} 6 & -1 & 3 \\ 4 & 0 & 2 \\ 1 & 8 & 12 \end{bmatrix} = \begin{bmatrix} 6x7 & -1x7 & 3x7 \\ 4x7 & 0x7 & 2x7 \\ 1x7 & 8x7 & 12x7 \end{bmatrix} = \begin{bmatrix} 42 & -7 & 21 \\ 28 & 0 & 14 \\ 7 & 56 & 84 \end{bmatrix}$$

A matrix can be multiplied not only by another matrix, but also by a single number, a scalar. This is achieved by the simple expedient of writing a second matrix of the same order as the first, but where every element is the product of the scalar and the corresponding element in the original matrix. Table 14 illustrates this generally and with respect to a numerical example.

Other forms of matrix arithmetic are less directly useful to the ecologist and will not be dealt with here. Good introductory accounts appear in Goult *et al.* (1973) and Shields (1968). The further applications of matrix algebra in ecology may be pursued in the work of Williamson, whose 1972 book is a readable and useful introduction to the topic.

5.3 Stability Theory

In 2.1, concerned with systems terminology, it was pointed out that frequently we wish to understand the dynamics of a complex system rather than its statics; that is to say, we wish to observe the behaviour of the state variables in the system through time. This concern with the dynamics of systems behaviour has led systems ecologists, in particular, into an area of analysis which can be referred to conveniently as stability theory. There is a mathematical theory of stability of complex systems, as recourse to any standard text on differential equations and related topics will show. At this introductory level, however, we need concern ourselves only with some of the basic ideas and terms involved. Consider a simple population model containing two species of animal which interact with each other, say a predator and its prey. The abundances of the two species will vary through time and, depending upon a number of other state or driving variables, a number of different sorts of behaviour might be observed.

The left-hand graphs in figure 18 show how some of these behaviours might appear when the numbers of the prey and the predator species are plotted through time. Graph a illustrates the situation where the prey is less than wholly suitable for the maintenance of the predator population and the predator species eventually disappears; graph b shows where the predation pressure imposed by the one species on the other is too great, leading to the disappearance of the prey species, and as a consequence of course the later demise of the predators also. Both these situations are intrinsically **unstable**; that is to say, the

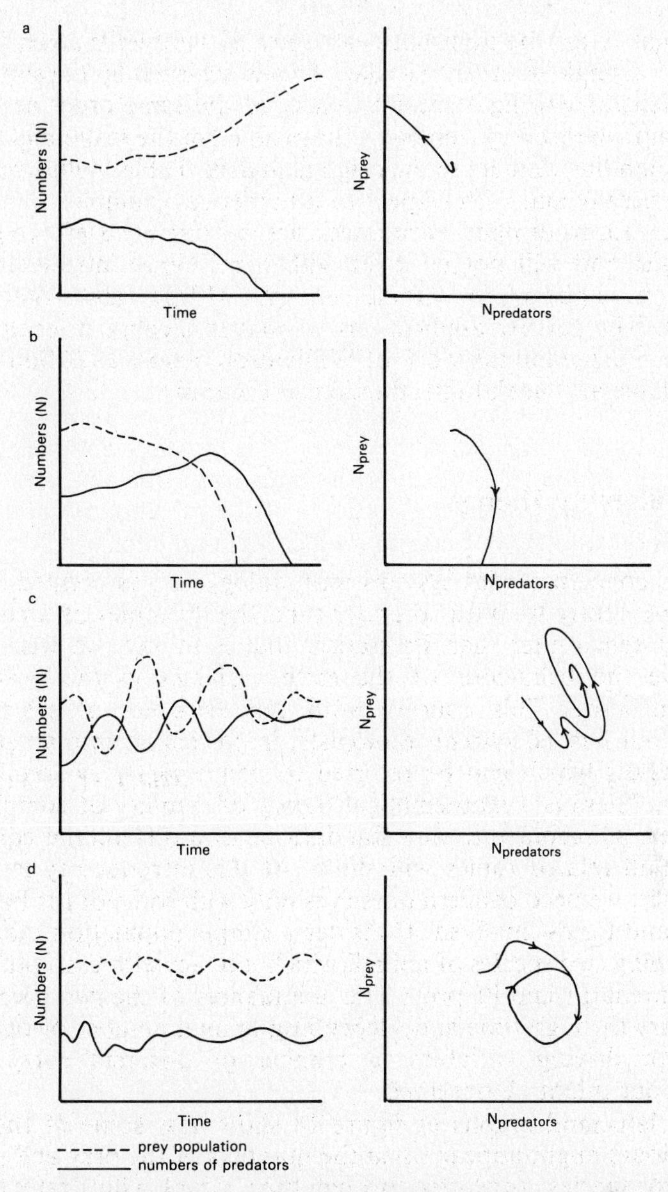

Fig. 18 Some predator-prey dynamics with differing stability properties. The left-hand curves represent the systems behaviour in the time domain, the right-hand ones, in the phase plane. a. Population of predators becomes extinct – the system is unstable. b. Population of prey becomes extinct – again an unstable system. c. An equilibrium is reached between the two populations after a period of oscillation – the system is stable and has an equilibrium point. d. Stable oscillations of the two populations are achieved after a time – the system is stable, exhibiting "limit-cycle" behaviour

two components of the simple system we are looking at cannot coexist indefinitely through time. Graph c shows a more viable situation where there is an initial period in which the two species undergo considerable fluctuation in numbers, but these fluctuations diminish through time until eventually a stable arrangement is reached where the mean numbers of each species remain more or less unchanged through time. In other words, both species oscillate in numbers but these oscillations are damped. Graph d in figure 18 shows an alternative form of stable arrangement where, following the initial fluctuations, the two species enter regular **cycles** of abundance, which can persist indefinitely.

A very useful way of looking at the stability or instability of the situations portrayed in the examples is by use of another type of graph, of the sort shown at the right-hand side of figure 18. To obtain these graphs the numbers of the predator are plotted against the numbers of prey at each point in time for which there is information, and in this manner plots are obtained in the **phase plane**. These points are joined up in the order in which they occurred in the time-based graph to obtain a **trajectory** showing the behaviour of the system. In the two unstable situations referred to, the trajectory resulting from a plot of this sort soon cuts one or other axis of the graph indicating that either the prey or the predator species has become extinct. In the third case where the predator and prey reach stable levels the phase plane plot shows a trajectory which spirals inward to a single point, the **equilibrium point**. In example d, the stable cyclic situation produces a phase plane plot which spirals into a circular trajectory indicative of what is known as a **limit cycle** situation. A number of other sorts of behaviour can be envisaged for systems of the sort described, but the ones referred to are those most useful in the interpretation of ecological systems.

So far of course, we have concerned ourselves only with a simple two-variable system for which clear and useful plots are possible in the phase plane (see figure 19). However, even in only moderately complex systems, many more variables are involved. The conceptualization of the behaviour of these variables with respect to one another is much more difficult than in the simple two-variable case but, in principle at least, they can be thought of in similar terms. With these more complex situations instead of a two-dimensional phase space, we must envisage an n-dimensional dynamical space, to use the terminology of Lewontin (1969), and the whole set of the

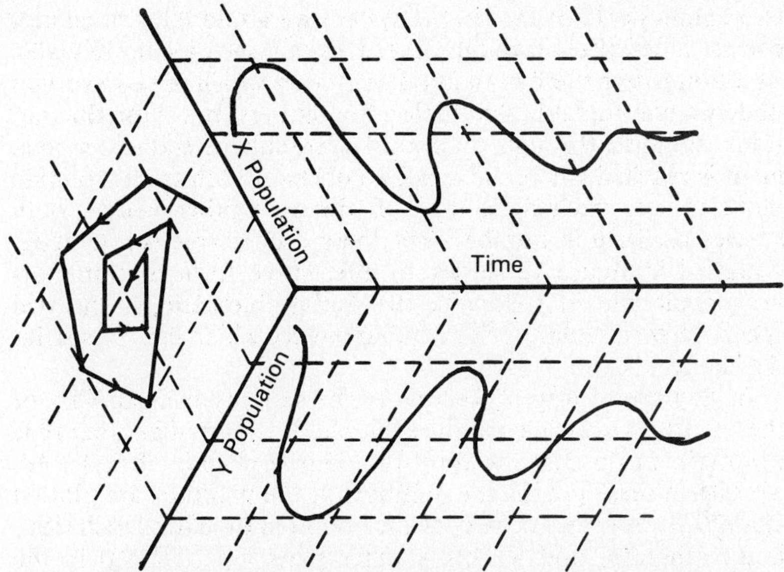

Fig. 19 The relationship between the time domain and the phase plane domain for a two-species system (after Holling 1973, reproduced, with permission from the *Annual Review of Ecology and Systematics,* Volume 4, © 1973 by Annual Reviews Inc.)

behaviours of a particular system can be reduced to descriptions of the behaviour of trajectories in this dynamical space. Let us return to the two-dimensional situation and explore some further concepts of stability, bearing in mind however that the concepts of which we speak can usually be generalized to the more realistic n-dimensional situation.

When considering the stability of an ecological system, we are concerned with the extent to which levels of a state variable can return to an equilibrium situation following a disturbance. We have already touched upon the different ways in which a return to stability can appear; we now need to consider under what circumstances such a return might occur. Some systems can be imagined, where the trajectory of systems behaviour will return to an equilibrium point or cycle regardless of how far from the equilibrium situation levels of the state variables are pushed. In graphical terms, this means that the trajectories in phase space will converge from any point in the phase space to a single equilibrium point in some instances or to the limit cycle situation in others. Such systems are said to be **globally** stable. In ecological systems, the much more common situation is where there is an area of the phase space surrounding the equilibrium point or cycle; from within this area the system will always return to the stable situation, and outside it the system will

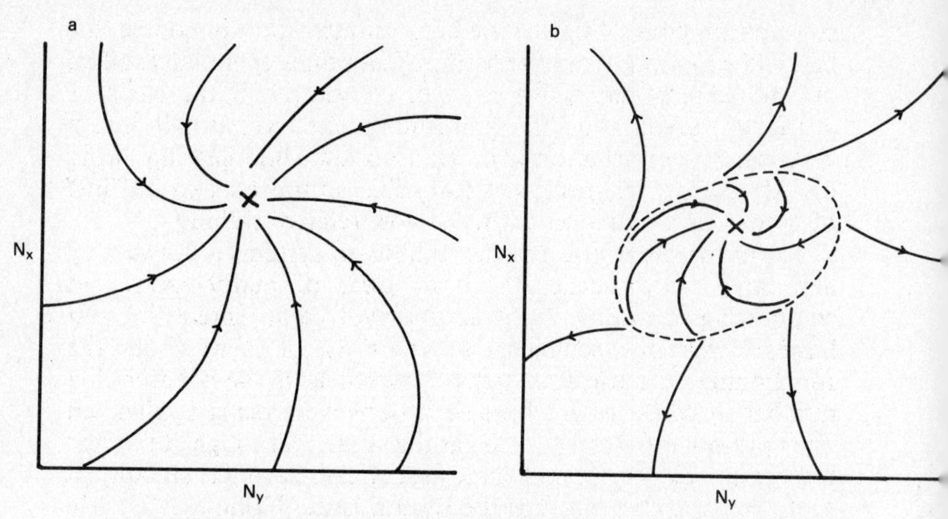

Fig. 20 Phase plane maps of two interacting populations (N_x and N_y), exhibiting (a) global stability and (b) neighbourhood stability. The dotted line delimits the domain of attraction of the system having neighbourhood stability

become unstable, with phase plane trajectories running outwards from the central area until they cut one or other of the edges of the graph. This is described as **neighbourhood** stability. Figure 20 illustrates the two notions of global and neighbourhood stability as they appear in phase plane diagrams. In the investigation and study of ecological systems, we are not so much concerned with whether stability is global or neighbourhood, as we can usually assume it will be neighbourhood, but with the size of the **domain of attraction** from within which stability is maintained. The size of this region gives a measure of the **relative** stability of the system, and permits us to compare systems describing different ecological situations, or of differing complexity but representing the same system, and so on.

Closely related to this is the speed with which a system will return to a stable state following a perturbation within the domain of attraction. In describing phase plane diagrams for the behaviour of systems, it is common to compare various trajectories in these diagrams with altitudinal contours on a two-dimensional map. Pursuing this analogy, the domain of attraction represents a basin and the speed with which a system will return to the stable state varies with the steepness of the sides of such a basin. We can imagine the system behaving like a ball-bearing placed on a contoured map: the steeper the sides of the basin, the sooner the ball-bearing will spiral down to the equilibrium point. Outside the basin of attraction of course, the ball will roll away from the equilibrium point, eventually falling off the edge of the universe involved. Lastly, if the basin of attraction surrounding an equilibrium point is very small, then a very small perturbation will lead to our ball-bearing being dislodged from the peak and following an unstable path. This, of course, is a situation with very low relative stability.

The above description of the stability of systems is derived, by and large, from ideas in use in areas of mathematics and engineering. Certainly the concepts involved are interesting and have led systems ecologists to some useful ideas about the functioning of natural systems. Nevertheless, the proliferating number of cases to which these notions were being applied led one prominent worker, to argue that, although of great mathematical interest and some ecological relevance, stability in ecological systems was not the central issue (Holling 1973). He suggested a separate although related concept, **resilience,** as the feature of most importance to the ecologist. The difference between stability and resilience can be summed up best perhaps in Holling's own words (p. 17): "Resilience determines the persistence of relationships within a system and is a measure of the ability of these systems to absorb changes of state variables, driving variables and parameters *and still persist* . . . . Stability on the other hand, is the ability of the system to *return to an equilibrium state* after a temporary disturbance" (my emphases). Holling reached this conclusion after observing that many natural populations fluctuated widely and irregularly through time but still persist. In other words, they are very resilient although not at all stable. His distinction ties in nicely with some of the recent applications of games theory to natural evolution. The process of adaptation and selection represents an on-going contest among species and between species and their environments, but a contest which no species ever "wins". Ultimately extinction is inevitable but the "aim" of the contest is simply to stay in the game as long as possible. Holling points out that, of course, there are links between stability and resilience and they can be regarded, in some way, as complementary concepts.

To return to our phase plan descriptions of systems behaviours, resilience is more concerned with properties of the **edge** of the domain of attraction than with its centre. Resilience measures, then, will be concerned with the overall area of the "domain of attraction" and the rates of "retreat", as it were, of trajectories from them. It is a view complementary to quantifying stability. We are concerned with the position of the equilibrium point or cycle and the rate of "return" to this equilibrium. In a personal communication Holling argues that resilience is a "boundary orientated" notion as opposed to the "point orientation" of stability. For the systems ecologist, interested in interpreting the behaviour of the models he build, Holling's ideas on resilience represent a useful addition to his armoury.

5.4 Probability

The equations and systems of equations encountered so far can be used to describe parts of the natural world. They contain numerical values which have been assumed to be constant; that is, they are **deterministic** systems (see chapter 2). In our observation of the real world, however, we will seldom if ever encounter quantities which are so invariable in time.

This notion can be further explained with a familiar example. Table 10 describes a series of age-specific birth rates for the four age classes identified in a population, and table 13 assigns numerical values of 0, 0.5, 2.0 and 1.0 to these parameters. Consider the birth rate for age class three. We have the statement that in every time interval, Δt, each individual gives rise to two offspring, no more and no less. Realistically, of course, this is absurd; sexuality apart, some individuals will have no offspring, some one, some two, some three and so on. However, we can have the situation where **on average** each individual has two offspring. We have now identified a probabilistic situation where there is a **distribution** of values around the mean. Figure 21 illustrates three situations, each of which results in a mean value of two. Histogram (a) is perhaps like our mental picture of how numbers "should" be distributed when we use a mean value by itself. The highest bar on the histogram (the "mode") lies on the mean value and the other bars diminish symmetrically on each side. The second case (b) shows a markedly different situation in which the mode lies to one side of the mean and the distribution is "pushed over" or **skewed** to the left when compared with the symmetrical situation. In addition, the values observed for brood size have a wider range than in the first case — 0 to 6 in comparison with 0 to 4. The third case is put in as a cautionary tale! Histogram (c) is symmetrical, mean brood size is again two, but in fact fewer parents had two offspring than any other brood size! Whereas the other distributions had a single peak, this one has two. It is said to be **bimodal,** contrasting with the **unimodal** situation displayed by distributions a and b. These distributions can be viewed in another way, by converting raw data to frequencies. This is done by expressing the number of instances of occurrence of a particular brood size as a proportion of the total number of observations made. For example, if data were collected on the brood size in thirty-eight individuals and four of these had five offspring, then the frequency of occurrence of broods of five would be 5/38, that is: 0.132. These frequencies show the same patterns of variation as the raw data but allow for comparability of information across samples of different sizes. The frequency of occurrence is a comparable number whether based on a total of sixty or six hundred observations. In addition, if we are satisfied that the distribution obtained is on the whole representative of the process, then we can move more readily from frequencies to **probabilities.** Thus, in histogram a of figure 21, the frequency of occurrence of broods of size two was 12/38, that is: 0.316. If we are satisfied that the shape of the histogram will change little if more data are collected, we can say that the probability of an individual producing a brood of two is 0.316. In other words, if we measured the brood size of one hundred individuals we would expect to find thirty-two of them to be of size two.

The shape, spread and skewness of a distribution can be calculated from the raw data using formulae developed by statisticians. These quantities are referred to as the **moments** of the distribution and some of the common computational equations are given in table 15. The derivation and theoretical background to these formulae can be found in most basic works of statistics such as Bulmer (1965) or Kendall and Stuart (1969). Knowing these measures describing the distribution about a mean value, we are in a much better position to appreciate the nature of what is involved at a particular system transition. We may be able to estimate the errors in our prediction of the new state vector after a transition, knowing the variance associated with our transition probabilities. However, errors will be cumulative and possibly multiplicative, and more than one transition precludes any accurate predictions concerning the errors involved.

Much of **parametric** statistics is based on the idea that measured quantities in the real world will conform to one of a few well-known **probability distributions**, and the mathematics

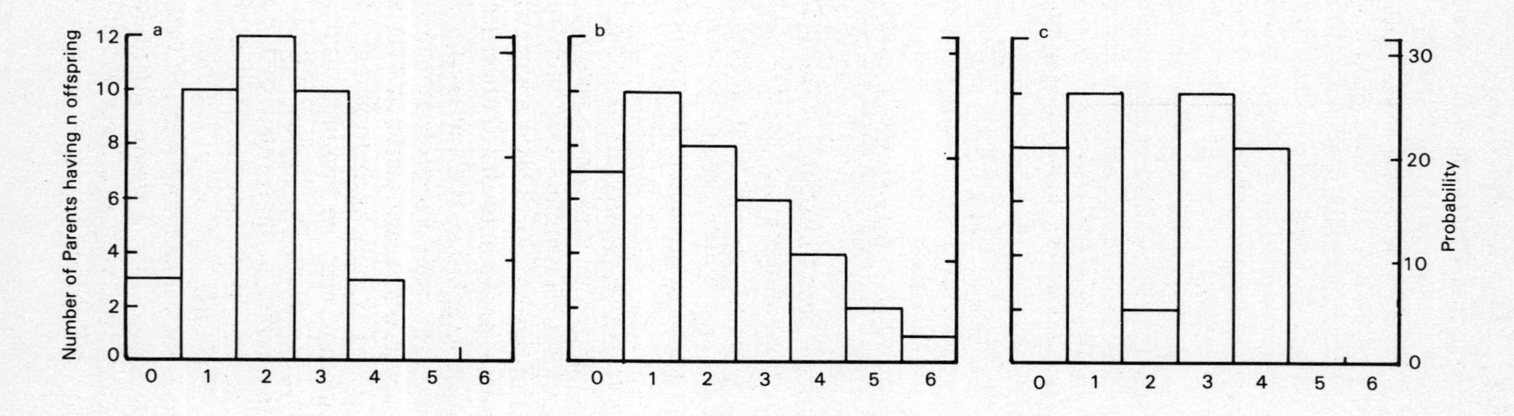

Fig. 21 Three distributions of reproductive rates, each having a mean value, over the whole population, of two, but representing a. a unimodal symmetrical, b. a unimodal skewed and c. a bimodal distribution

Table 15. Some statistical formulae for the description of distributions (based on Bulmer 1965)

For a series of observations x_i where $i = 1$ to n the following moments can be computed.

The Mean ($\bar{x}$) or average is a measure of the centre of a distribution.

$$\bar{x} = \sum_{i=1}^{n} x_i/n$$

The Variance (m_2) is a measure of the spread of a distribution.

$$m_2 = \frac{\sum_{i=1}^{n} (x_i - \bar{x})^2}{n - 1}$$

The -1 in the denominator is a correction for relatively small sample sizes.

The Skewness (m_3) is a measure of the shape of a distribution, being zero for the symmetrical case.

$$m_3 = \frac{\sum_{i=1}^{n} (x_i - \bar{x})^3/n}{(\sqrt{m_2})^3}$$

Other formulae for skewness are given and explained in Bulmer (1965).

---

of these distributions have been thoroughly investigated. Three such distributions of particular interest in the present context are shown in figure 22. The case illustrated in graph a, usually referred to as the **uniform** distribution, is more a reference point than a frequently used case. It implies that there is an equal probability of occurrence of all values within the range of the variable concerned. A truly random process in principle has this sort of distribution and the "pseudo"-random numbers generated by compilers of most computers, suitably ranked and plotted, will conform to this pattern. This is a somewhat trivial example of a **continuous** probability distribution (see Bulmer 1965, p. 36). In contrast, graph b shows a **discrete** distribution, but one which results, also, from a random process. If the individuals of a plant species, say, are scattered over an area at random, that is, if they are equally likely to lie anywhere in the area, then the population may be sampled using standard quadrat methods (see, for example, Greig-Smith 1964). This involves counting the number of individuals of the species of interest within each quadrat. If we plot the number of quadrats containing 0, 1, 2, 3 plants, and so on, then we obtain a distribution which may look something like the one shown in graph b. This is referred to as a **Poisson** distribution and is characterized statistically by having a variance which is more or less equal to the mean value. Many ecological variables conform

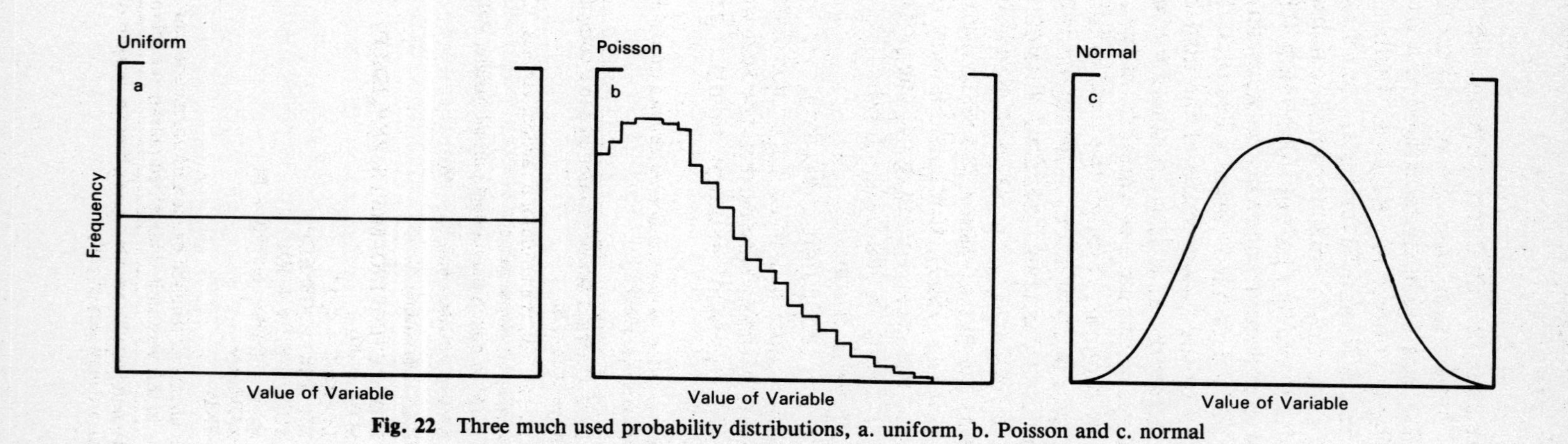

Fig. 22 Three much used probability distributions, a. uniform, b. Poisson and c. normal

to this or related distributions, although very frequently the variance is larger than the mean, indicating a degree of clumpedness or **contagion** characteristic of a grouping of points in space. These notions are explored further by Southwood (1966), Pielou (1977) and Seber (1973), among others. Graph c of figure 22 is the most familiar symmetrical distribution, of the kind alluded to in the first part of this section. This is the *normal* or Gaussian distribution, showing a characteristic bell-shaped curve when plotted. It has a variance less than the mean and the mean is equal to the modal value of the distribution. Like the uniform case, graph a, it is a continuous distribution and is exemplified by a vast range of variables in the real world, from the heights of humans, through the head-widths of midges, to gene frequencies in populations. The degree of conformation of a set of data to a particular standard distribution, such as the Poisson or normal cases, can be evaluated using conventional statistical tests, details of which can be found in works such as Kendall and Stuart (1969), Cox and Hinkley (1974) or Gilbert (1973). It is to the great advantage of the putative modeller to identify such forms in his data, if possible, as these can be simulated readily in his model.

Frequently in simulation modelling, we wish to generate values drawn at random from a particular distribution in order to mimic some quantity observed in the real world. Most compilers contain "pseudo"-random number generators, which

Table 16. The generation of random normal deviates from uniform random deviates (after Gordon 1969)

If we required random normal deviates conforming to a normal distribution of mean, M, and standard deviation, S:

1. call a sequence of, say, twelve uniform random variates, RX;
2. sum these variates to obtain SUM;
3. operate on SUM to obtain the random normal variate, RNV, using the formula

$$RNV = [(SUM - 6.0).S] + M$$

or, as a FORTRAN subroutine:

```
      SUBROUTINE INORM(M, S, RNV, ISEED)
      SUM = 0.0
      DO 100 I = 1, 12
      RX = RAND(ISEED)*
100   SUM = SUM + RX
      RNV = (SUM - 6.0)* S + M
      RETURN
      END
```

where ISEED is an integer required by the uniform random number generator to "seed" its operation – in this version it is initialized outside the subroutine programme.

* Note that different installations may have different versions of this intrinsic function and methods of use may vary slightly.

produce numbers on call drawn from a uniform distribution; these random **variates** can then be manipulated to produce variates conforming to a particular function. The most frequently needed case is that of the normal distribution. A very simple method of obtaining random normal variates, using a summation procedure operating on a number of uniform random variates, is laid out in table 16, together with the FORTRAN code required to implement it as a programme subroutine. Gordon (1969) discusses the operation of this and other similar transformations in detail.

6

Population Systems

This is the first of three chapters dealing with a series of examples of systems ecological studies, which illustrate many of the principles and methodologies already mentioned. This chapter concentrates on a series of case histories where the primary level of study has been the **population**, chapter 7 looks at ecological processes operating and modelled at the **individual** level, and chapter 8 considers the **community** and **ecosystem** level of organization and their representation in quantitative terms. It is worth reiterating, however, that these levels are chosen for convenience only, both by the workers involved and in my descriptions of their work. In any particular study, minor and major excursions are made across level boundaries, and I shall make similar jumps whenever necessary.

The population level has been chosen as the introduction to this sequence of examples, mainly because "population" has played such a central role in the development of ecological theory and practice since the inception of the subject at the beginning of this century. This is particularly true with respect to animal ecology, although in recent years extensions of the approach into plant ecology have been made by workers such as Harper and his school (see, for example, Harper 1977). In addition to this central place in ecological theory, the population is also the level of organization of natural systems that first attracted the attention of bio-mathematicians, such as Lotka and Volterra, and it was perhaps only natural that many of the early excursions into systems ecology should have involved similar areas of study.

6.1 Population Dynamics

For the reader unfamiliar with the basic tenets and principles of population dynamics, I shall briefly review the concepts involved. This is necessarily an introductory treatment and the interested student is referred to more extensive and exhaustive accounts in such texts as Solomon (1969), Wilson and Bossert (1971) and Emmel (1976).

The study of population dynamics is concerned with the **changes** that occur in the quality and quantity of individuals within a biological population through time. There are only four basic ways in which the number of organisms in a population can change in any time period. Additions to the population can be made in either or both of two ways: through births or by immigration from surrounding populations. Similarly, reduction in numbers can occur by the two complementary processes: death and emigration from within the particular population to the surrounding populations. The identification of these four basic processes — **natality, immigration, mortality** and **emigration** — leads us to the basic equation of change in population dynamics, which represents a summation of these four quantities. Table 17 summarizes some of the elementary mathematical relationships in population dynamics, the first being this basic equation of change. In much of the formal theory of population dynamics, the processes of immigration and emigration have been ignored, assumed to be either negligible or else equal to each other and thus mutually cancelling. In this elementary theory, the number of births and the numbers of deaths in any period is considered to be a function of the

Table 17. Basic equations of population dynamics

1. Population change in any time interval = Births (B) + Immigrants (I) - Deaths (D) - Emigrants (E)

2. (a) $B = bN$
   (b) $D = mN$
   where N is population size at any time, b is the birth **rate** and m is the death **rate**.

3. $$\frac{dN}{dt} = bN - mN$$
   $$= (b - m)N$$
   If (b - m) is defined as the intrinsic rate of increase, r, then

4. $$\frac{dN}{dt} = rN$$

5. If the environment is finite with a carrying capacity K, then
   $$\frac{dN}{dt} = rN(1 - \frac{N}{K})$$
   where (1 - N/K) represents a restraint on the growth rate related to how close the population size is to the carrying capacity of the environment.

number of animals actually occurring in the population and, in the simplest possible case, this relationship is assumed to be linear. Equations 2a and 2b in table 17 show how the number of births and deaths can be related to the numbers in a population by the simple inclusion of constants, the instantaneous birth rate and death rate. Considering these two factors alone, we can proceed from the simple rate equations to a basic differential equation representing change in numbers of the population through time. This is equation 3 in table 17, which shows the rate of change of the population as related to the numbers in that population at any point in time and the difference between the birth and death rates of the population. This difference between the birth and death rate is usually called the instrinsic rate of increase and is conventionally designated, r.

This substitution gives us the equation of change labelled 4 in table 17, which should be recognizable as the differential equation depicting unrestricted, geometric or exponential growth of a population that we encountered in chapter 5. As noted earlier, this represents adequately the growth of a natural population only under special sets of circumstances. It assumes a superabundance of the resources the population needs, and as such is only representative of newly establishing populations, either natural or artificial. The characteristic curves representing this sort of accelerating growth have been obtained and published for stored products beetles (Hardman 1976a), pheasants (Einarsen 1945) and even human beings (for an early account see Malthus 1798). Nevertheless, this geometric model of population increase is hardly realistic for most natural situations, and one step up in the move towards a realistic model is achieved by the logistic equation. Again, the form of this equation is explained in chapter 5, the key addition being a "carrying capacity" term, conventionally designated K, such that the rate of increase of the population diminishes as its number approaches this carrying capacity. The equation involved is equation 5 in table 17. Once again this is a simplified version of what one would expect to see in most natural populations, but curves approximating that described by this formula have been published for a variety of organisms, including fruit flies (Pearl 1925), *Paramoecium* (Gause 1934) and even fur seals (Kenyon and Schaffer 1954). Like the exponential model, the logistic equation most satisfactorily describes the changes in population levels following the invasion of a previously unoccupied area. Both of these models make a number of unrealistic assumptions about the organisms they describe and the environments in which these organisms live. In systems terminology, their most serious shortcoming is their assumption that the organisms concerned operate in closed systems, that is, that there are no influences acting upon the population other than those of the population itself. The birth and death rates are assumed to be constant through time, the age structure of the population is supposed to be stable and unvarying, the environment in which the organisms exist is assumed homogenous both in space and time and, associated with the closed nature of the system, there are no significant driving variables acting upon the population.

All of the factors mentioned above and others, mean that there has been considerable difficulty in applying simple algebraic models of the kind described to anything other than very selected examples. Some workers have chosen to circumvent this difficulty by largely abandoning the use of such simple basic models in favour of more complex models, such as the one described for aphids in 6.4. Some authors, however, have attempted to use the idea of exponential growth as the basis for more realistic and flexible models, and have incorporated features into their models to attempt to overcome some of the shortcomings of the algebraic expressions. Two pieces of work in this category will be described in 6.2 and 6.3. The first was carried out by the Canadian systems ecologist, Michael Hardman, who made extensive studies of the population dynamics of the beetles which infest stored grain in Australia. He constructed a sequence of models based on both the exponential and logistic formulae incorporating age structures into the models, in both deterministic and stochastic versions. He made comparisons between these models and data collected in real situations. The models he built of exponential or, as he calls it, Malthusian growth will be described shortly. The second example chosen builds upon the matrix representation of the age structure and dynamics of populations already alluded to in chapter 5. The work of a variety of authors will be described, building upon the basic theory of Leslie (1945, 1948).

Before proceeding to the examination of these particular examples, however, it is necessary to cover briefly one other aspect of population dynamics, which has special application in many studies of cold-blooded animals, which frequently form the subjects for modelling efforts. Population dynamics is the study of the changes in populations or organisms through **time**; there is some question, however, of just how to interpret the notion of time when considering cold-blooded animals. The metabolic processes of such animals are to a greater or lesser extent dependent upon the temperature of their environment. Thus, for example, for an insect a hot day may represent a greater length of subjective time than a cool day, providing the opportunity for more development and more activity and, accordingly, for the completion of those ecological processes which depend upon the activity and general physiology of the animal. In building models of population dynamics of insects, for example, special account has to be taken of this **temperature** dependence. It is no good producing models which predict population dynamics in terms of days and hours when nothing is known about the amount of heat input to the animal occurring during this period of calendar time. To cope with this problem, entomologists in particular have used the concept of **day degrees**. In essence a day degree is a unit of temperature plus an associated time scale. If we measure the rate of development of an organism at a variety of temperatures, we obtain a series of developmental rates; that is, the reciprocal of the developmental time, and a plot of these rates against temperature produces, in the simple linear case, a graph such as that shown in figure 23. From a graph of this sort we can identify the temperature below which development is negligible — the **developmental zero** or threshold temperature.

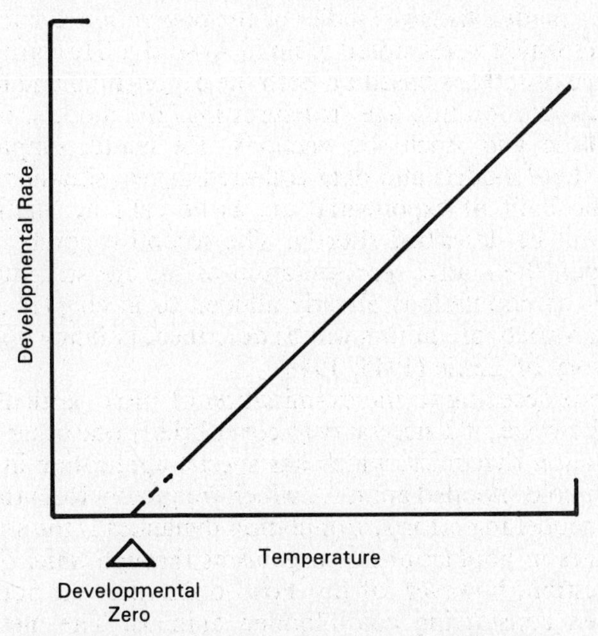

Fig. 23 Notional plot of the developmental rate of a poikilotherm and the way of calculating the developmental threshold temperature

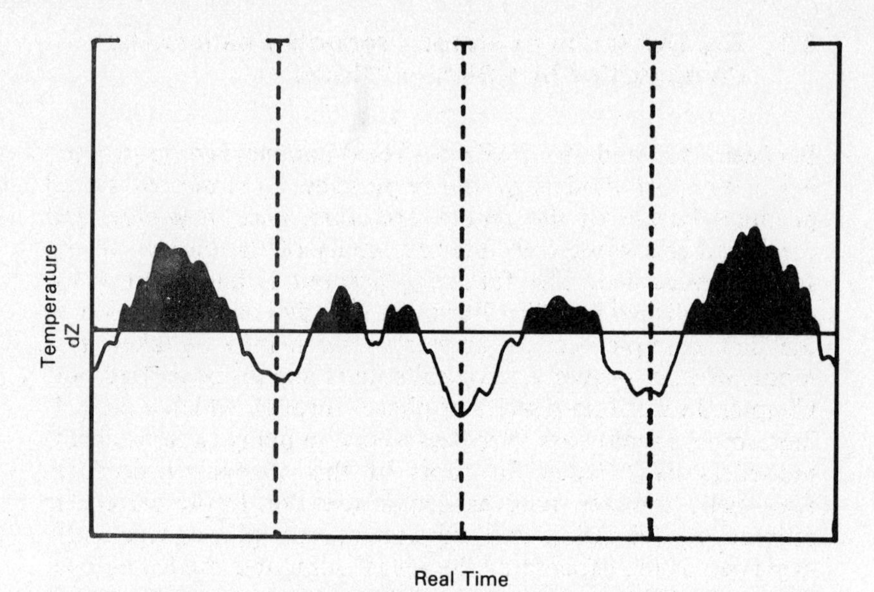

Fig. 24 The concept of thermal summation. The number of day degrees elapsing in any period is found by integrating beneath the daily curve of temperature, above the developmental threshold of temperature (dZ).

Knowing this temperature, we can obtain field data on ambient temperatures in the region of interest and superimpose the developmental zero upon these data in the manner shown in figure 24. Integrating below the temperature curve for the period of interest and above the line representing the development zero, we obtain a measure of the **developmental** or **physiological time** experienced by the organism in the field during that period. The unit involved is the day degree. In any associated modelling exercise, then, we use temperatures collected in the field as a driving variable determining the rate of development experienced by the animal on any particular day, hour, week and so on. Kitching (1977) provides an extended account of this concept and describes some of the complications which may occur when, for instance, the relationship between temperature and time is non-linear, when the normally available shade temperatures cannot be used and so on. This work also discusses development, temperature and reproductive activity in insects with particular reference to blowflies. More recently, Taylor (1981) has produced a more theoretical treatment on "physiological" time in insects and its evolution, referring to its practical consequences as time-"warping".

6.2 The Dynamics of Stored Products Beetles: The Complexities of a Simple "Model"

Between 1972 and 1977, Michael Hardman worked on producing a series of models of the population dynamics of stored products beetles of the genera *Tribolium* and *Sitophilus*, and combined this work with intensive study of the biology of the species concerned. The following account is based upon two papers published in 1976. Hardman's study has been chosen as the first example because in many ways it may be taken as a model of how systems ecological studies **should** be carried out. Chapter 3 identified a series of phases through which a modelling project should pass. Needless to say, in many circumstances, modellers have fallen far short of this stepwise procedure. Hardman, however, paid particular attention to the parameter estimation, validation and experimentation phases frequently neglected at the expense of the actual computer modelling procedure. In addition, he described his work in lucid terms, and his model demonstrates by this good example how necessary it is for the systems ecologist to develop substantial biological insight with respect to the organisms and their environment that are the subjects of the modelling activities. The model produced is then a useful tool rather than a house of cards based on little or no real information.

The practice of storing wheat and other grains in silos before marketing is widespread, and a large variety of pest organisms have become adapted to exploit these stores of grain. Flour beetles of various sorts are among the most prominent and economically important of these species, and considerable effort has been expended in studies of their population biology (see, for example, Birch 1948, Howe 1952, Mertz 1969, Niven 1967, and papers by Park and his associates, 1948, 1965). Some of these authors, including Birch (1948) and Deevey (1947), have used the simple model of geometrical increase already mentioned in order to describe the population growth of beetles following an invasion of such bulks of stored grain. Hardman recognized that this very simple analytical model is restrictive because of the many assumptions it makes about the biology of the animal and its environment. His inital aim, realized in this model, was to build simulations of geometric growth. Subsequently, he built a more realistic model, this time of logistic growth of beetle populations which incorporated more detail about the biology of the species and the environment in which it occurs. This account has been restricted to Hardman's models of geometric growth, but will describe these in considerable detail as an appropriate introduction to the whole modelling business.

Hardman built three models of geometric growth, of increasing complexity. In the first, which he referred to as his **first deterministic model**, all of the biological parameters were obtained by averaging techniques from his observations of the biology of the subject species. His second and third models aimed to circumvent the oversimplifications caused by this averaging, by introducing increasing degrees of stochasticity into updated versions of the first model. All of these models divide the population into a number of age classes — the egg, larvae, pupae, adults and "old" adults, which latter class Hardman referred to as "seniles". All three models, in addition, are designed to follow the changes in numbers of particular age classes in the population through time, and sum these in order to mimic the behaviour of the population as a whole.

The **first deterministic model** is summarized in the flowchart given in figure 25. Like most models of population dynamics it is an iterative one, that is, the passage of time is represented by repeated passage through a series of calculations within a time loop. The amount of temperature-mediated physiological time which is assumed to have occurred during each pass through the time loop is a variable quantity, which can be changed at the whim of the modeller either from run to run of the model, or from time to time within a single run of the model. This model assumes that the growing population consists of discrete **cohorts**, or sets of individuals, each such cohort being the result of the reproduction occurring within a particular passage through the time loop. Bearing these two basic decisions in mind, we can work through the structure of the model in some detail.

Before any calculations can proceed, certain values have to be read into the model to summarize the population biology of the species. These are the so-called **life table parameters**, and include the durations of each of the age classes in terms of the mean number of day degrees required to pass through that particular stage, the mean reproductive rates for each age class, and the mean value for the sex ratio in the population. Hardman obtained values for all of these quantities from experimental work carried out separately from the modelling exercise. He was deeply involved in the laboratory and field estimation of such parameters, in addition to being the prime mover in the modelling exercise. Following this input of information, at the beginn-

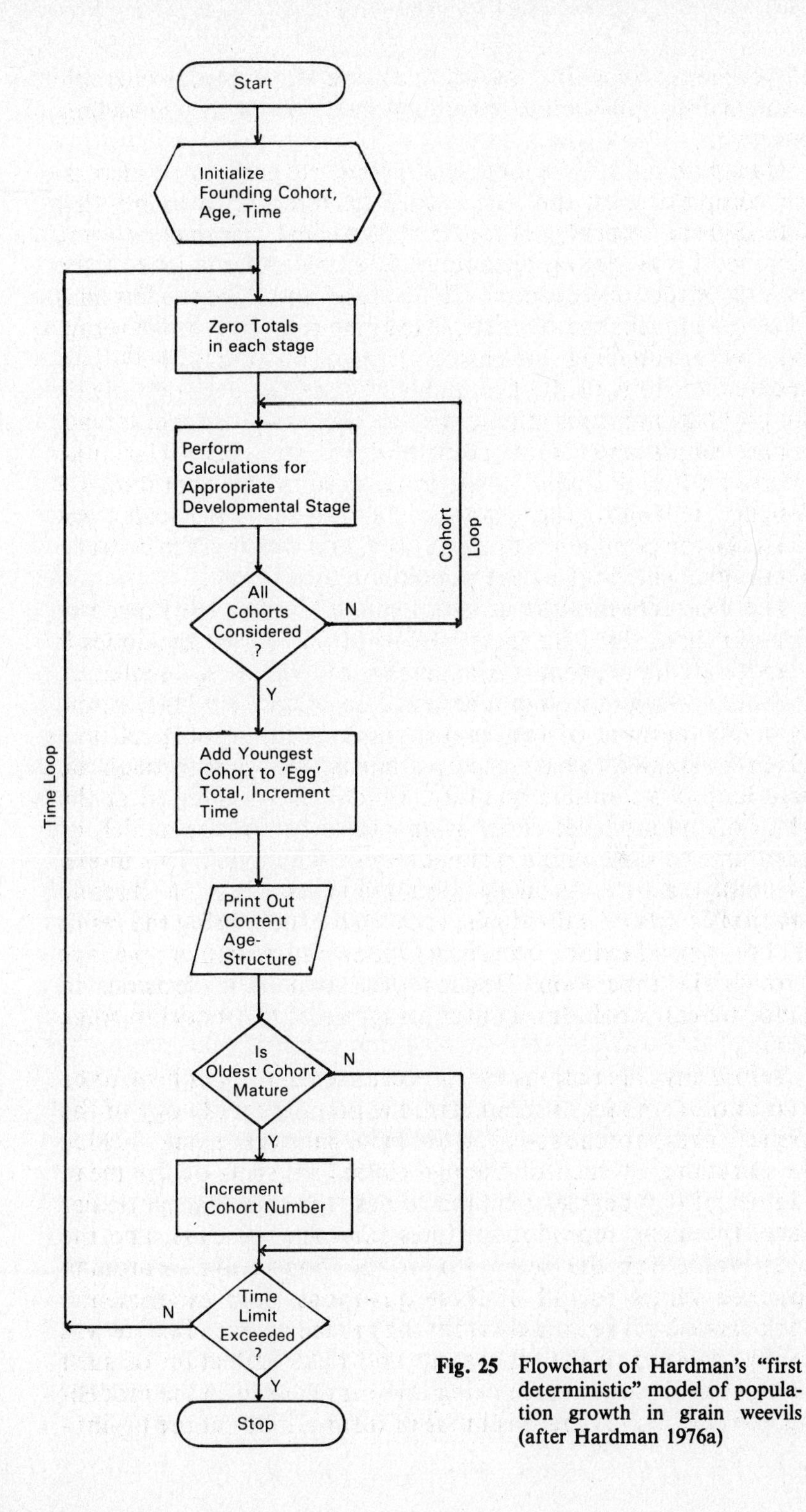

Fig. 25 Flowchart of Hardman's "first deterministic" model of population growth in grain weevils (after Hardman 1976a)

ing of any run of the model an initializing procedure has to be completed. This involves defining the number of individuals in each age class at the beginning of the sequence of calculations; these figures represent the first cohort to be considered by the model. The next and subsequent cohorts are generated within the model each time it passes through the basic time loop. In practice, the initial conditions of this founding cohort were defined to be such that all its members were the same age at the same stage of development. As in many similar models, this is purely a device of convenience; many possible age structures could be used to start the model running. Also during this initializing phase the running totals of cumulative real time and physiological time are set to zero.

The computational flow through the model then enters the major time loop. Each passage through this time loop mimics the effects of the passage of the predetermined amount of physiological time, and the model is designed to estimate the amount of **development, reproduction** and **mortality** that takes place during this period. Firstly, within the time loop, the running totals for each age class of the population are set to zero. This is necessary as the remainder of the calculations within the time loop are designed to update the totals for each stage, giving the age distribution at the end of the period of time assumed to have elapsed within the loop. The chain of calculations then enters a second loop within the larger time loop, and this considers each cohort of animals within the population being studied and performs particular calculations on each, depending upon the developmental stage involved. The model tests the age of the cohort being considered at the particular point in time and, on the basis of the results of this test, carries out a sequence of calculations appropriate to the stage involved. Figure 26 shows the flowchart for these "within-cohort" calculations. The calculations aim to simulate the action of three ecological processes acting upon the organisms within each cohort. The first is the **developmental process,** by which animals can move from one age class to the next. If such a move is deemed to have taken place then the totals of animals in the two classes involved in the move within the cohort must be adjusted. This is achieved by the mechanisms shown as the first stage in the flowchart for larvae, pupae, adults and seniles. It is not carried through for the egg stage as entry to the egg class is controlled by the calculations representing the reproductive adults. The only other mechanism which must be included in the chain of calculations is that which imposes a **mortality** upon the animals within that

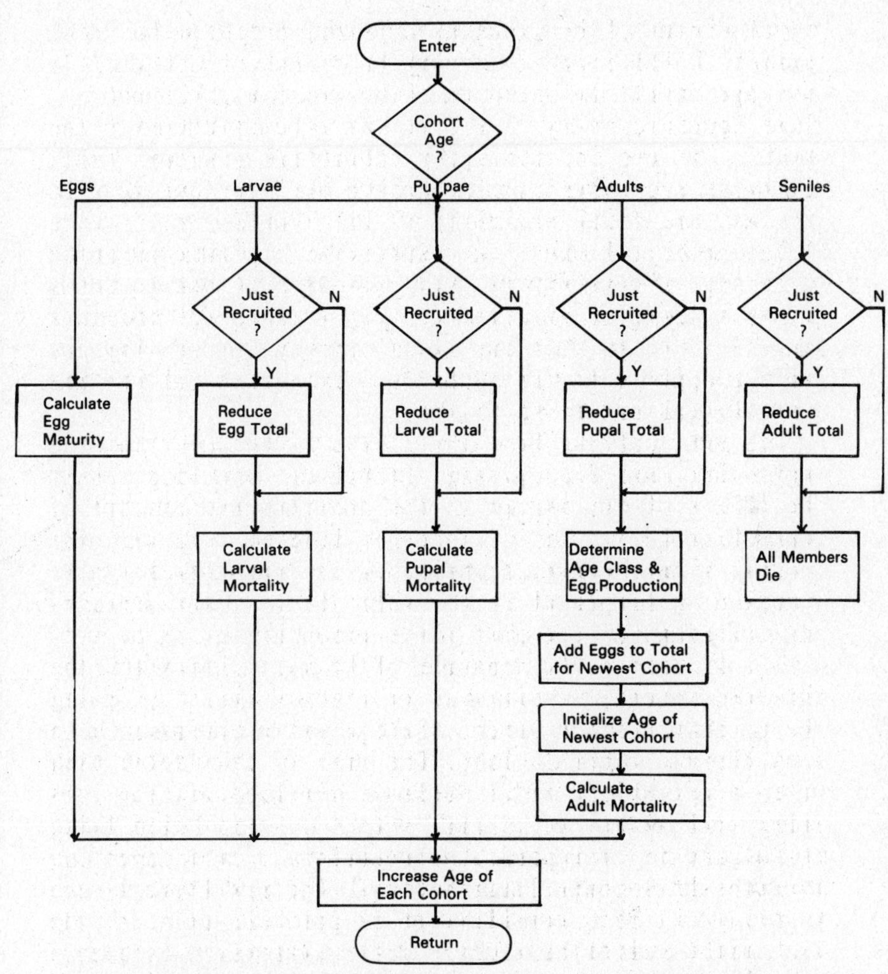

Fig. 26 **Flowchart of the "within cohort" calculations of Hardman's "first deterministic" model (after Hardman 1976a)**

age class. Hardman uses the same mechanism for calculating this mortality in all stages except the seniles, for which he makes the assumption that mortality will be one hundred per cent in any time interval. For the other age classes, a nice combination of biology and mathematics is used for calculating just how many animals should be considered to have died in that time period. The theory, laid out in table 18, is based on the idea that we can measure the number of animals that die within any one stage and work out a mean value for this within-stage mortality. Hardman did this in laboratory trials in parallel with his modelling activities. However, the time step used within any

Table 18. The theory of the calculation of mortality in each iteration of Hardman's "first deterministic" model

For each stage the total within-stage mortality is **measured** and is assumed to follow a negative exponential pattern. Thus the number surviving at the end of the step $N(t + \Delta t)$ is obtained from the numbers at the beginning of the stage $N(t)$ and is calculated from

$$N(t + \Delta t) = N(t)\, e^{-D} \qquad (1)$$

where D is characteristic for the stage and calculated knowing the overall mortality within the stage.
If the total duration of the stage is now thought of as comprising n time steps then

$$N(t + \Delta t) = N(t)\, e^{-nd} \qquad (2)$$

where d is the characteristic exponential for one time step and e^{-d} is the survival rate per time step.
If a day degrees pass within an interation of the model and e^{-d} is calculated as the survival rate per day degree, then the rate for that time step is e^{-ad}.

passage through the time loop of the model is unlikely to coincide precisely with the overall mean duration of any particular stage; even if it did coincide for one stage, it would not necessarily coincide for any of the others. What we need therefore is a method to work out how to modify the mortality rate known to apply over the whole age class, in order to obtain the rate for segments of that stage. Using information available from a wide range of other studies on insect survival, Hardman was able to assume that mortality within any one stage would follow a negative exponential curve. This was a reasonable assumption, especially as survival curves of this sort have been obtained for flour beetles under a great variety of circumstances by other authors. Having made this assumption, it is possible to represent the change in population from the beginning of the stage to the end by equation 1 in table 18. We know the numbers surviving at the end of the stage and we know the numbers at the beginning of the stage; we can then calculate D in that equation where D is the characteristic exponent for the equation representing survival within the stage and e^{-D} is the survival rate. If we know the total duration of the stage in smaller time steps, in hours, days, day degrees or any other recognizable unit of time, we can calculate survival **per unit time** by dividing the characteristic exponent just calculated by the number of time steps we wish to consider. This is because we can substitute in equation 1 to obtain the form shown in equation 2 of table 18.

Returning now to the model, if a certain number of day degrees are known to have elapsed during the current passage through the time loop, and if we have calculated our survival rate per day degree, then we can compute the survival rate for

Table 19. An example of the calculation of mortality in each iteration of Hardman's "first deterministic" model

---

If we know that the egg stage of *Tribolium* undergoes a 23 per cent mortality over its whole duration, we can write the survival equation for one hundred animals over the whole period as follows (see theory in table 18):

$$100e^{-D} = 77$$

from whence we can calculate D

$$D = \ln\left(\frac{1}{.77}\right) = 0.261$$

Now if we have measured the duration of the egg stage and find on average that it lasts twenty day degrees, we can calculate the mortality rate **per day degree** as follows:

$$e^{-D} = e^{-nd}$$

i.e. $e^{-0.261} = e^{-20d}$

hence $d = \dfrac{0.261}{20} = 0.01305$

and the mortality rate is $e^{-0.01305}$ per day degree.

If the iteration includes twelve day degrees then the mortality rate will be $e^{-(12 \times 0.01305)}$

i.e. $e^{-0.1566}$

and the basic transition equation will be

$$N(t + \Delta t) = N(t).e^{-0.1566}$$

---

the whole of that passage through the time loop simply by multiplying the exponent by the number of day degrees. This is explained in the last part of table 18. Knowing the survival rate, we can easily obtain mortality rates if required, as mortality is simply the complement of survival. Table 19 shows a numerical example of this sort of calculation, and explains in detail just what arithmetical manipulations are needed in order to apply Hardman's method of calculation.

Interspersed between the development and mortality process in the adult stage is the **reproductive process.** The computational method involved is set out in table 20. Hardman calculated the

Table 20. The reproduction process in each iteration of Hardman's "first deterministic" model

---

If the sex ratio (that is, the proportion of females) is S, the duration of the iteration is a day degrees and of the whole adult stage, A day degrees, the total adult fecundity (that is, eggs per individual female) is F, and the current population of adults is N(adults) then the number of eggs produced within the iteration, N(eggs), will be obtained from:

$$N(eggs) = [N(adults) \times S \times F] \frac{a}{A}$$

For example if N(adults) = 100
S = 0.5
F = 80
a = 12
A = 20

Then $N(eggs) = (100 \times 0.5 \times 80) \dfrac{12}{20} = 240.$

---

number of eggs laid in any time period from the number of adults present at that time, their sex ratio, and the age-specific fecundity of the adult stage, together with knowledge of the duration of the adult stage in day degrees. The fecundity figures and the duration figures, of course, were obtained from his observations on the biology of the organisms in the laboratory. Eggs deemed to be newly laid within each passage through the time loop form the initiating individuals of a new cohort in the model, which will age and go through all the associated population processes in the next and all subsequent iterations of the model. For programming reasons, it is necessary to keep account of the number of cohorts involved at any time as this determines how many passes the model should make through the "cohort" on the next passage through the time loop. Accordingly the updating of this total is the last process involved within the time loop. It is carried out quite simply by determining whether or not any of the cohorts presently in the model have reached maturity, that is the state of egg production. If they have, this means that some eggs must have been "laid" during the previous calculation and a new cohort initiated. The total of cohorts can then be incremented by one. Of course, if no cohort has yet reached the stage of egg production then the number of cohorts being considered by the model would be the same the next time through the time loop.

This completes the calculations involved within the time loop, and the model repeats these for the next time around. The number of passages made by the model around this loop is entirely an operational decision on the part of the modeller; in practice, the modeller frequently aims at running the model for a sufficient length of time so that any peculiarities associated with the arbitrary initialization procedures are submerged in the subsequent dynamics of the model.

A careful examination of the structure of this model will show that it has most of the drawbacks of the simple algebraic model for exponential population growth. It does, however, contain more biological structure in so far as it incorporates age-specific developmental rates, mortality rates and reproductive rates. It has been described in some detail because it contains many features characteristic of a large class of simulation models of populations, and is sufficiently simple to serve as a useful and clear introduction to the methodologies involved. Hardman himself was well aware of the shortcomings of this first model. He was mainly concerned with its **deterministic** nature; that is to say, mean values were used for the biological parameters involved such as sex ratio and the various rates processes. This is obviously unrealistic, as each such parameter has an associated probability distribution, and the occurrence of any particular value in the real world would be probabilistic rather than fixed. He built two further models in a partial attempt to overcome these shortcomings. In this account, I propose only to mention the principles involved in his changes, rather than go into the model structures; his own papers can be consulted if detailed information on the specific computational changes is required.

The only major change in Hardman's second model was to incorporate the observed variability in the durations of developmental stages into the calculations. From observations of the real world, Hardman knew how these durations varied and was able to represent them as probability distributions. In the model this change is implemented in the developmental process within the cohort loop, and this means that, when a particular cohort reaches a given age, it is not transferred piecemeal into the next developmental stage. Only a proportion of individuals are passed on, depending upon the age of the cohort, and the exact number considered to have matured is determined from the experimentally determined probability distributions of stage durations. All other parameters in the model use mean values as before.

Hardman described his last model of Malthusian population growth, as his **stochastic model.** It is a great deal more complicated than previous models and, in addition to having a probabilistic representation of the duration of developmental stages, also incorporates similar variability in sex ratios, mortality and fecundity rates. This requires a fundamental change in model structure, as it is necessary to keep track, within the model, of the progress of individuals rather than of cohorts. Decisions about whether particular individuals proceed to the next stage, die or give birth, in the particular time period, are made using a random process operating upon the known probability distributions surrounding the mean values for these parameters. Following the reproductive process, the sex of a particular animal is assigned using a similar probabilistic procedure. Such random processes are based on the availability of random number generators within the compiler of the computer, and in this regard Hardman follows methods similar to those described in 5.4.

In terms of the nine stages of the ecological modelling procedure identified in chapter 3, the above account has covered Hardman's progress through problem definition, systems identification, his decisions as to the structure, mathematics and nature of the model to be built and the programming procedures. His work with respect to the next three stages in the procedure are, however, worthy of a little more consideration.

We have alluded peripherally to the procedures used for **parameter estimation**. Initially, Hardman thought that sufficient information about the biology of flour beetles existed in published form to provide adequate information for input to the model. As it turned out, however, there was a great deal of disagreement among authors as to what were characteristic values of the various life table parameters for the beetles concerned, so Hardman found it necessary to study the biology of the beetles in detail in parallel with the modelling process already described. The methods and results of this process are the subject of a separate paper published at the same time as that describing his modelling activities (1976a). Hardman reared his insects at a variety of temperatures in order to obtain precise durations for each of the age classes in which he was interested; he observed pairs of beetles in order to obtain the fecundity schedules required for the reproductive process incorporated into his model; he determined rates of mortality during the course of his rearing experiments to provide the survivorship curve incorporated into the model; and he examined the sex ratio, again at a variety of temperatures. All of these observations he was able to summarize statistically as means and frequency distributions. Indeed, in many respects Hardman's model is a great deal more satisfactory than most, in so far as he was able to use parameter values in which some confidence could be placed and was not reduced to any of the makeshift or guessing procedures alluded to earlier as the unavoidable lot of many modellers.

The **validation** phase was also given a great deal of attention. Experimental populations of beetles were established in glass containers where the physical environment closely resembled the silos in which infestations of the subject species occur. These were suitably replicated and seeded, in the one instance with one hundred eggs of the beetles, and in the second case with eighty adults. Each was sampled over a fifty day period and estimates of the numbers of each age class present at each sampling time obtained. Hardman attempted to validate and make comparisons among the three models described by using the values of the driving variables, temperature in particular, that pertained in his aquarium trials, and thereby predicting the population dynamics that would occur in them according to each of his

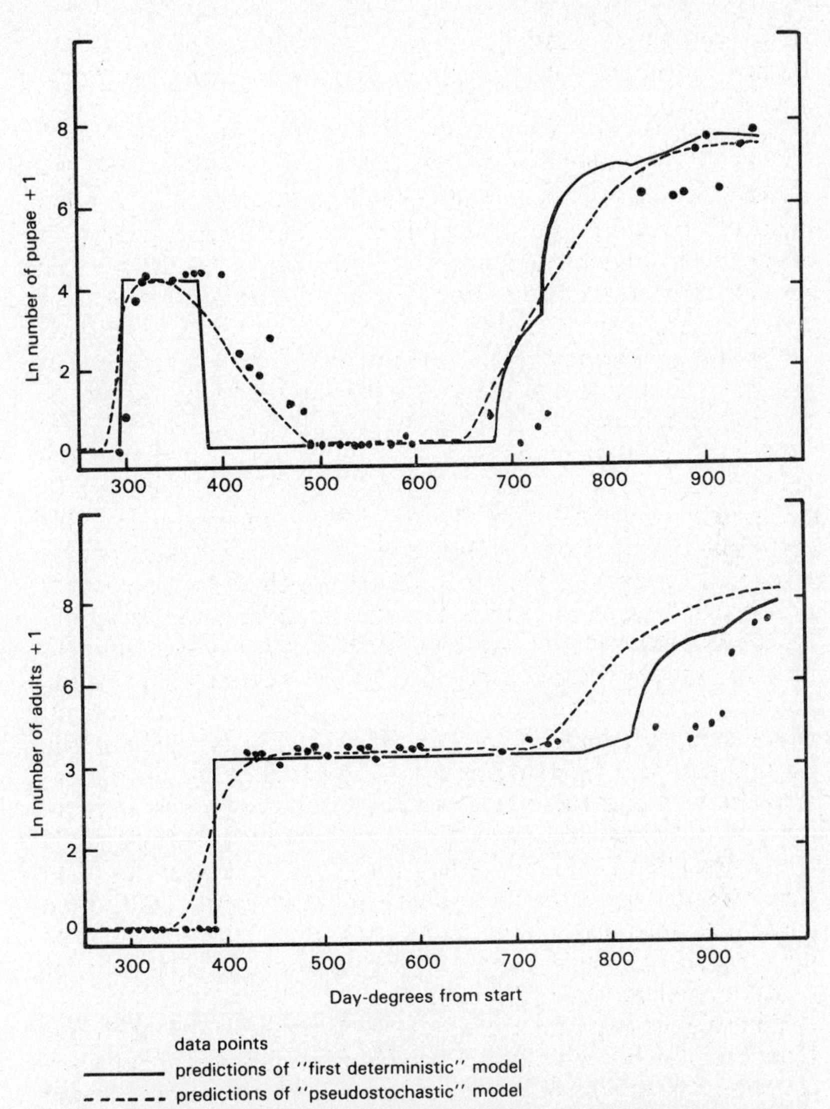

Fig. 27 Comparison of actual and predicted populations of pupae and adults of flour beetles, *Tribolium confusum*, in aquaria (after Hardman 1976a)

three models. The output from these runs was then compared with the actual population dynamics observed in the experimental situation. Figure 27 compares the actual population dynamics observed in Hardman's trials with the predictions obtained from running, firstly, the simple deterministic model and, secondly, the model which incorporated developmental stochasticity into the computations. The third and last model that Hardman built, the so-called stochastic model which

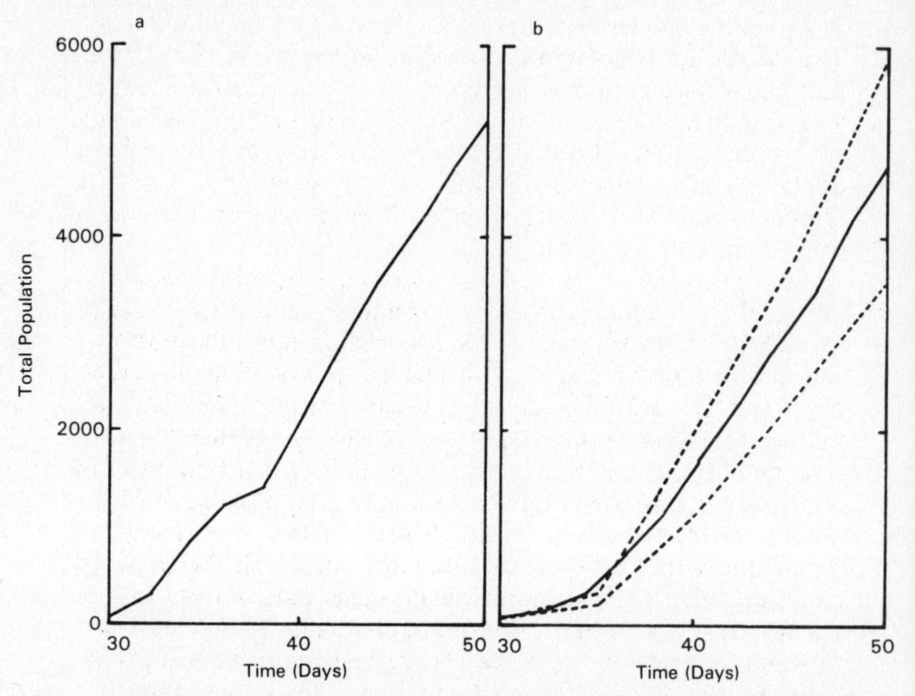

Fig. 28 a. Growth of a population of *Tribolium confusum* reared under Malthusian conditions at 30°C, 70 per cent r.h. as predicted by Hardman's "deterministic" model. The starting population was one hundred unsexed 0-24 hour old eggs. b. Growth of the population under the same conditions as a. Solid line is the prediction of the pseudostochastic model, while upper and lower broken lines are the maximum and minimum populations out of ten replicates generated by the "stochastic" model (after Hardman 1976a)

introduced variability into all the life table parameters involved, proved very hard to validate in this fashion. The reason was that the individual nature of the computations meant that obtaining a computer run simulating a lengthy period of time required a great deal more machine space and time than was available, and hence a coarser form of comparison was all that was possible. Figure 28 shows this, comparing as it does the growth of the total population of beetles predicted by the stochastic model over the first fifty days from establishment, with a similar prediction of the two earlier models. Comparison of the curves predicted by the models with the data shows that in most respects the models Hardman built were good predictors of the changes that occurred in real populations of flour beetles. The second model, which has some stochastic elements in it, gave a better fit to both the pattern of changes observed in the population and the magnitude of the numbers involved, although the first two models overestimated both adult and pupal numbers. The validation process for the third, stochastic model showed that the model predicted the growth of the overall population with fair accuracy, but it also indicated that the great variability in numbers that Hardman noted in comparing the results obtained from one aquarium to the next was similarly duplicated by the predictions from different runs of the simulation model.

Hardman went on to use his model to make predictions about the results of manipulations of parameter values. He chose to investigate three separate processes in this fashion: the mortality rates, duration of stages, and the fecundity schedule. With respect to the mortality rates, Hardman was interested in determining the relative importance of mortality operating at different points in the life cycle of the beetle. Thus, instead of applying the experimentally determined mortality rates in his model, he made runs in which these were changed, and compared the output of these manipulated runs with the standard run using the experimentally determined parameters. In this fashion he was able to show that high larval mortality has a very substantial effect on the overall success of the population. That is to say, the models he built had a high sensitivity to the value used for the larval mortality. A number of biological reasons could be forwarded for this, one of which is the longer duration of the larval period. Without a complex model such as the one Hardman built, it is unlikely that this feature of the biology of the animal would have been identified as of particular importance. Hardman's second set of experimental runs related to the duration of the developmental period. He pointed out that various factors have been recorded which may retard the developmental rate of stored products beetles, such as low relative humidity, low food quality or deterioration of the grain bulk as a result of previous insect activity. In order to investigate what effects such factors might have on the population dynamics of the species, he manipulated the values for the durations of particular life history stages that were entered into the model, and followed the effects on the model's predictions. Once again he was able to show that the model was most sensitive to changes made in the larval stage, and that increasing the length of this stage had a profound effect upon numbers of both adults and immatures. Lastly, in this experimental phase of his work, Hardman looked at the effects of reduced fecundity on the population dynamics, as simulated by his models. Yet again he was able to point out that reduced fecundity was observed in natural situations following sub-lethal doses of pesticides, shortage of oviposition sites, overcrowding, fouling and undercrowding, as well as adverse temperature and humidity conditions. He simulated this effect by reducing the fecundity level across all age classes to 50 per cent and 20 per cent of their normal values. Again, as might be expected, this had significant effects on population size and a 50 per cent reduction in fecundity led to a 41.6 per cent reduction in population size after seventy days of population growth at 30°C. Even further depressions in population size resulted from the reduction to 20 per cent of the normal fecundity values.

There are several general observations to be made about the class of models of which Hardman's is representative, and which can be described most generally as **book-keeping models.** Hardman's sequence of constructs ran into a difficulty common to many such models. The addition of even moderate amounts of complexity, as in his third model, used an excessive amount of computer storage and time in order to operate on anything like realistic levels. This feature of book-keeping models of the kind described is a result of their mathematical simplicity. In order to introduce more components into such a model more structure is added, rather than a streamlining of the mathematics, and problems frequently result from the cumbersome nature of these additional structural elements. Nevertheless, and off-setting this disadvantage, the models remain of considerable value, especially as heuristic devices, because they are easily understood by the non-mathematician. It is a common experience of systems ecologists however, that having built and worked with models of this sort, one soon looks to the next stage in mathematical sophistication. One very flexible, useful and powerful mathematical device which permits this upgrading is the use of **matrix representation** of populations and their dynamics. Cuff and Hardman (1980) rewrote Hardman's (1978) logistic model of population growth — itself an extension of the Malthusian models described here — using such techniques. A series of further examples using matrix methods of computation comprises the next section of this chapter.

6.3 Blue Whales, Red Deer and Great Tits: Matrix Models of Animal Populations

Hardman built models to represent the dynamics of animal populations by incorporating information gathered about the life history of the animals into a model which kept track of groups of animals through time. This book-keeping approach was built up from an initial consideration of simple models of population growth written in differential form, the exponential and logistic relationships. Chapter 5 showed that an analytical approach to the representation of population dynamics can go further than these simple differential equations models. The representation of population changes using matrices permits many of the operations that Hardman incorporated into his book-keeping model to be followed in a more mathematically elegant manner, and this sort of model will be explored in this section.

The great potential for the application of matrices in population modelling was recognized by P.H. Leslie in 1948. He presented the basic algebra involved, and a great many other workers have subsequently elaborated on his methods and provided specific examples of their application in natural, managed and artificial populations. An excellent introduction to the applications of matrices in population ecology is provided by Williamson (1972) and Usher (1972), and Enright and Ogden (1979) review some of the recent developments in the field.

There was in fact, a Leslie matrix in the examples dealt with in 5.2. To recapitulate, a Leslie matrix is one which has the age-specific fecundities across the top line and the age-specific mortality rates down the principal sub-diagonal. Such a matrix is show in table 21, where the elements f_0, f_1, and so on

Table 21. Leslie matrix – basic form

$$\mathbf{L} = \begin{bmatrix} f_0 & f_1 & f_2 & \cdots\cdots & f_{m-1} & f_n \\ p_0 & 0 & 0 & \cdots\cdots & 0 & 0 \\ 0 & p_1 & 0 & \cdots\cdots & 0 & 0 \\ 0 & 0 & p_2 & \cdots\cdots & 0 & 0 \\ \vdots & \vdots & \vdots & & \vdots & \vdots \\ 0 & 0 & 0 & \cdots\cdots & p_{n-1} & 0 \end{bmatrix}$$

represent the average number of births per individual in any time period, and the elements p_0, p_1, and so on represent the probability that an individual of the age represented by the first subscript will survive into the age class represented by the next subscript in any period of time. To be precise, these elements usually refer to the females in the population only although, with only minor modification, as we shall see later, we can treat with both sexes in any population. The matrix represents the transition mechanism by which the population ages in any time interval and, as we have seen already, we can write the time to time transition as follows:

$$\mathbf{n}_{t+\Delta t} = \mathbf{L}\mathbf{n}_t$$

where **n** represents the vector of age classes in the population at any point in time. We can carry this one step further and, by a simple extension to the methods of matrix algebra that we met in chapter 5, we can show that, after k such transitions, the relationship between the present age structure and original age structure will be represented by the following:

$$\mathbf{n}_{t+k\Delta t} = \mathbf{L}^k\mathbf{n}_t$$

where k is the number of time intervals and all other symbols are as before.

Usher (1972, 1976) has applied this sort of model to populations of the blue whale (*Balaenoptera musculus*). Using data collected by Ehrenfeld (1970) and Laws (1962), he was able to estimate values for the elements of the matrix. He divided the whale population into seven age classes, each of two years' duration, with the last class dealing with individuals in their thirteenth and all subsequent years. Perhaps not surprisingly, there is little concrete information on the mortality rate and fecundity schedule for whales, and the sources that Usher used in determining what figures to put into his model were themselves based on somewhat hypothetical relationships. Ehrenfeld presented a fecundity versus age curve for the blue whale, which incorporated information about age at sexual maturity for females, gestation period, nursing period and period between pregnancies. Ehrenfeld himself points out that the curve he presents shows maximum possible fecundity for each age class: however, he does claim that it is fairly accurate over the initial steeply rising portion of the curve. Laws presented estimates of survival with age for a related species of whale, the fin whale (*B. physalus*), showing that they had a survival rate which was more or less constant through time, that is, the mortality rate was more or less constant for all ages of whales. The fecundity curve and the survivorship curve on which Usher based his models are presented in figure 29. Using these sources of information, Usher was able to construct the Leslie matrix given in table 22.

To run this simplest of Leslie matrix models, Usher initialized the vector representing the population of whales, setting the youngest age class and the fourth age class, that is the six to seven year olds, to a hundred each and all other age classes as zero. Using the matrix described in table 22 and the basic transi-

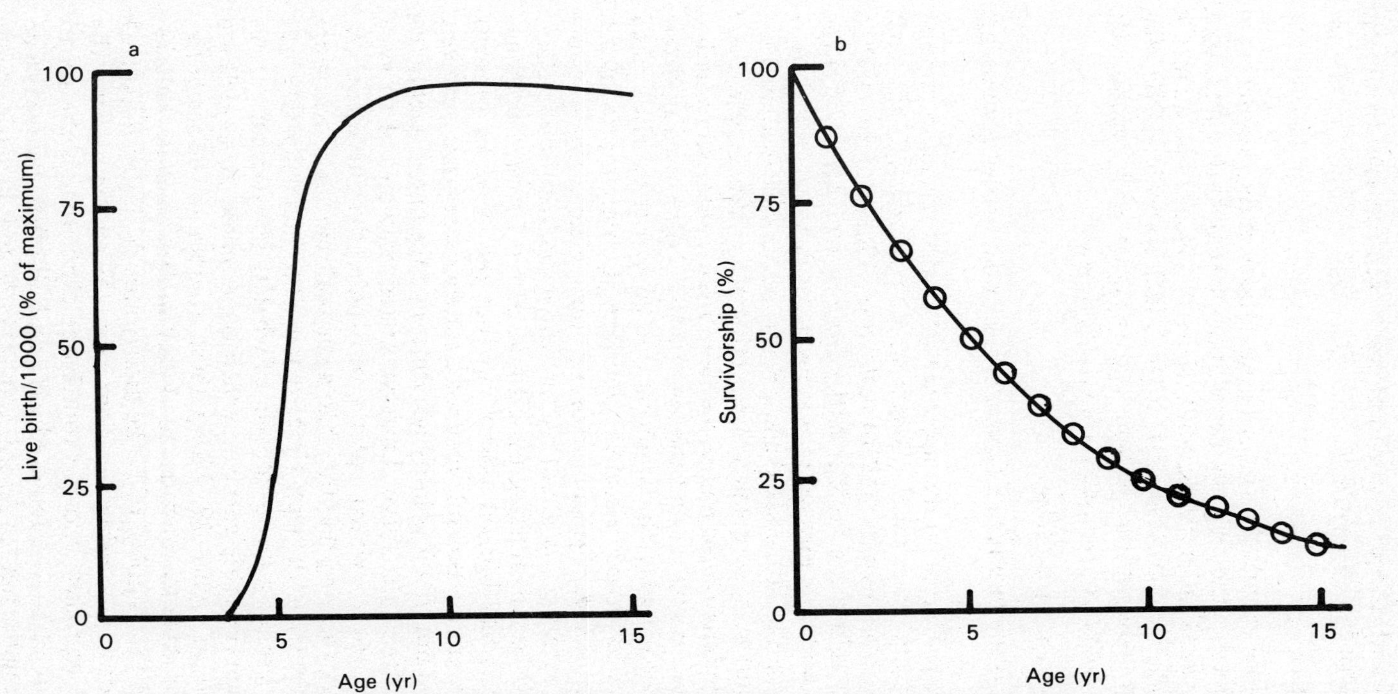

Fig. 29 Curves of a. reproductive rate and b. survivorship used by Usher (1972) in his model of blue whale populations. Curve a is based on data of Ehrenfeld (1970) and curve b on those of Laws (1962) for the related fin whale

Table 22. Leslie matrix constructed by Usher for the blue whale

$$\mathbf{L} = \begin{bmatrix} 0 & 0 & 0.19 & 0.44 & 0.50 & 0.50 & 0.45 \\ 0.87 & 0 & 0 & 0 & 0 & 0 & 0 \\ 0 & 0.87 & 0 & 0 & 0 & 0 & 0 \\ 0 & 0 & 0.87 & 0 & 0 & 0 & 0 \\ 0 & 0 & 0 & 0.87 & 0 & 0 & 0 \\ 0 & 0 & 0 & 0 & 0.87 & 0 & 0 \\ 0 & 0 & 0 & 0 & 0 & 0.87 & 0.80 \end{bmatrix}$$

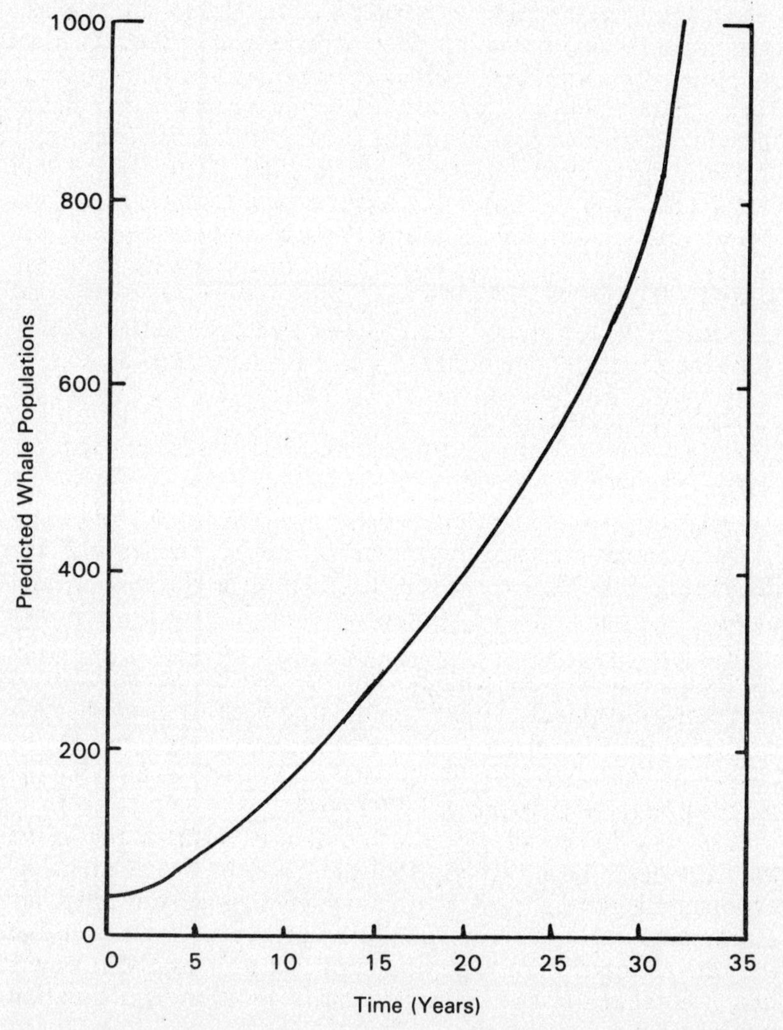

Fig. 30 Curve predicting levels of populations of blue whales obtained using Usher's 1972 model

tion equation for the Leslie matrix, he was able to produce a series of predictions as to how the total whale population would change and what its age structure would be. Figure 30 shows some of the results he obtained, drawn from data presented in his 1972 paper. He observed that after thirty applications of the transition multiplication the age structure, represented on a proportional basis, was stabilized. This is to say that, although the numbers in each age class continued to increase, the proportional representation of each age class in the population as a whole did not. Any subsequent applications of the transition matrix failed to change this situation. The factor of increase between the thirtieth and thirty-first application of the transition matrix can be obtained from the ratio of the two totals involved. This is 3608/3284, that is 1.0987. It will be observed that all the elements in the vector of the age structure at the thirty-first iteration are 1.0987 times the corresponding element of the vector following the thirtieth iteration. In mathematical terms, this constant multiplier is referred to as an **eigen value** of the matrix, and a considerable body of theory exists in linear mathematics dealing with the behaviour and properties of such numbers. The vector representing the age classes in the population after the achievement of the stable age structure just alluded to, is called an **eigen vector** of the matrix. We can write this property of the model for this stage as:

$$\mathbf{Ln} = \lambda \mathbf{n}$$

where λ is the eigen value and $\mathbf{n}$ is the eigen vector. λ will be one if the population is stable; that is, neither increasing nor decreasing; less than one if the population is decreasing; and more than one if the population is increasing. Further multiplication by λ will produce more eigen vectors, in unending sequence. All of these can be summarized by reduction to a single proportionality vector by dividing each element through by the value of the element having the greatest magnitude in any particular vector. Table 23 illustrates these notions from the results presented in Usher's 1976 paper.

It is easy to see when we arrive at this form of the Leslie matrix model that it differs very little from the simple model of exponential growth such as that which formed the basis for Hardman's model described earlier. The λ value is a measure of the rate of increase of the population. In fact, as Usher points out, it is related to the "r" of the exponential growth equation by the simple expression

$$r = \ln \lambda$$

Table 23. Eigen values and eigen vectors from Usher's 1976 model of blue whale dynamics

From an initial population vector (100 0 0 100 0 0 0) multiplication by the Leslie matrix:

$$\begin{bmatrix} 0 & 0 & 0.19 & 0.44 & 0.50 & 0.50 & 0.45 \\ 0.87 & 0 & 0 & 0 & 0 & 0 & 0 \\ 0 & 0.87 & 0 & 0 & 0 & 0 & 0 \\ 0 & 0 & 0.87 & 0 & 0 & 0 & 0 \\ 0 & 0 & 0 & 0.08 & 0 & 0 & 0 \\ 0 & 0 & 0 & 0 & 0.87 & 0 & 0 \\ 0 & 0 & 0 & 0 & 0 & 0.87 & 0.80 \end{bmatrix}$$

produces the following sequence of age vectors from the thirtieth application of the transition process onwards:

$$\begin{bmatrix} 725 \\ 575 \\ 455 \\ 360 \\ 285 \\ 226 \\ 658 \end{bmatrix} \rightarrow \begin{bmatrix} 797 \\ 631 \\ 500 \\ 396 \\ 313 \\ 248 \\ 723 \end{bmatrix} \rightarrow \begin{bmatrix} 875 \\ 693 \\ 549 \\ 435 \\ 345 \\ 272 \\ 794 \end{bmatrix} \rightarrow \begin{bmatrix} 961 \\ 761 \\ 603 \\ 478 \\ 379 \\ 299 \\ 872 \end{bmatrix} \rightarrow \text{etc.}$$

The age structure is now stable, each element being 1.0987 times the corresponding one in the preceding vector. This is the **eigen value** and those vectors are **eigen vectors.** We can obtain **a standard form** by dividing through by the magnitude of element 1, whence we obtain (when rewritten as a row vector):

(1 .792 .627 .497 .393 .311 .907)

where 1n indicates use of a natural (base e) logarithm. It can be used to obtain an estimate of this intrinsic rate of increase (r), and for the blue whales that were the subject of Usher's model this came out as 0.0940. In the form described, the matrix model has the same limitations as the simple algebraic model of geometric growth and, like that model, it is an adequate description of the growing phase of a population when the environment is more or less unlimited, but becomes inadequate in the more common natural situation where population numbers are such that resource limitation is a real factor. The shortcomings of the model can be described under a number of heads:

1. the Leslie matrix describes only the female dynamics;
2. the survival terms and fecundity factors in the matrix are considered to be unchanging through time and do not respond to changes in the environment or the biology of the species concerned; and
3. although the survival terms in the model are based on probabilistic notions of mortality, the model is in fact deterministic as only single values are used for both fecundity and survival.

A number of extensions of this approach to the modelling of populations have been developed, and are the subject of the reviews by Usher (1972) and Enright and Ogden (1979). Two such extensions are dealt with briefly here — that which introduces males into the model and that which replaces the deterministic values in the model with functional relationships. The third drawback of the model, its determinism, has been overcome, in theory at least, by the work of Pollard (1966), who has developed a stochastic form of the basic model just described, although it is highly technical and inappropriate for an introductory treatment such as this.

The incorporation of both sexes into the Leslie matrix model requires only a very simple extension of the original form of the matrices and vectors involved. The extension was first put foward by Williamson (1959) and the transition relationship involved takes the form given in table 24.

Table 24. Williamson's extension of the Leslie matrix for two sexes

$$\begin{bmatrix} 0 & f_{mo} & 0 & f_{m1} & 0 & \cdots & 0 & f_{mn-1} & 0 & f_{mn} \\ 0 & f_{fo} & 0 & f_{f1} & 0 & \cdots & 0 & f_{fm-1} & 0 & f_{fn} \\ p_{mo} & 0 & 0 & 0 & 0 & \cdots & 0 & 0 & 0 & 0 \\ 0 & p_{fo} & 0 & 0 & 0 & \cdots & 0 & 0 & 0 & 0 \\ 0 & 0 & p_{m1} & 0 & 0 & \cdots & 0 & 0 & 0 & 0 \\ \vdots & \vdots & \vdots & \vdots & \vdots & & \vdots & \vdots & \vdots & \vdots \\ 0 & 0 & 0 & 0 & 0 & \cdots & p_{mn} & 0 & 0 & 0 \\ 0 & 0 & 0 & 0 & 0 & \cdots & 0 & p_{fn} & 0 & 0 \end{bmatrix} \begin{bmatrix} n(t)_{mo} \\ n(t)_{fo} \\ n(t)_{m1} \\ n(t)_{f1} \\ \vdots \\ n(t)_{mn} \\ n(t)_{fn} \end{bmatrix} = \begin{bmatrix} n(t+\Delta t)_{mo} \\ n(t+\Delta t)_{fo} \\ n(t+\Delta t)_{m1} \\ n(t+\Delta t)_{f1} \\ \vdots \\ n(t+\Delta t)_{mn} \\ n(t+\Delta t)_{fn} \end{bmatrix}$$

The fundamental difference between this matrix and the straight-forward one-sex version described earlier is that it is bigger. In fact, it is twice the size in both dimensions. There are now two rows representing fecundity, each of which has half of its elements set to zero as the males, by definition, have zero fecundity at all ages. The non-zero elements in the two rows represent the age-specific production of male offspring, in the first row, and female offspring, in the second. The rest of the matrix comprises two rows for each age class, the first referring to males and the second to females, and once again probabilities of survival form a diagonal through the matrix. This doubling of the size of the transition matrix has, implicit in it, a doubling of the size of the population vector. This now comprises two entries for each age class, the first referring to the males and the second to the females.

There have been few actual applications of this model since its introduction by Williamson in 1959; for this reason I turn to yet another model constructed by Usher, this time referring to the population dynamics of red deer *(Cervus elaphus)* and based on data collected by V.P.W. Lowe (1969). Lowe obtained data which permitted the estimation of age-specific survival terms for each of the sixteen year classes into which he divided the deer population. The age-specific fecundity terms for males and females were rather more difficult to obtain from Lowe's data, but Usher was able to use the available information on the proportion of the population that was breeding at any time to obtain estimates of the required fecundity rates. A 32 × 32 matrix was obtained and the figures involved can be obtained by reference to Usher's works (1972, 1976). As with the blue whale population, iteration of this model enabled Usher to obtain a value for λ which, once more, was positive (in fact it equalled 1.1636), and from this information he was able to obtain a standard eigen vector representing the stable age structure.

Usher comments at length, with respect to both blue whale and red deer models, on the possibilities of using the information gleaned from the modelling exercises as a basis for cropping practices on the two species. He points out that the rate of increase obtained by repetitive multiplication and represented by the eigen values in both cases can be used as a guide for the proportion of a population which can be harvested in any period of time. In the case of the red deer, he points out that this is not an unreasonable estimate as Lowe's data refer to an unexploited population on the Scottish island of Rhum. In the case of the whales of course, the fact that the biological parameters involved are derived from very rough estimates of what is going on in what is already a heavily exploited population, makes it doubtful if the maximum exploitation rate which can be inferred from the λ values obtained from his model could be sustained. In both instances it is assumed that exploitation takes members of all age classes indiscriminately and, of course, this is unlikely to be true. However, as Usher himself points out, these sorts of models can be used to estimate the effects of removal on specific age or sex classes.

The replacement of the elements of the Leslie matrix by functional relationships is described in papers by C.J. and L. Pennycuick (Pennycuick, Compton and Beckingham 1968, Pennycuick 1969), using data on a bird, the great tit (*Parus major*), collected by a variety of workers in the Oxford area of England. These data were reviewed by David Lack in his 1966 book.

The model that the Pennycuicks built, although using Leslie-type transitions as its basic computational device, has a great deal more structure in it than previous matrix models that we have considered and, in many ways, represents an approach intermediate between the simpler Leslie matrix models and the "life system" models to be described in 6.4. The necessity of computing mortality and fecundity terms required that the model take on a form much more reminiscent of the numerical simulations of Hardman than the concise, deterministic, matrix models of Usher. This model is based on the premise that fecundity and survival in the great tit can be identified with particular times of the year. Pennycuick (1969) examined the large body of data available and identified relationships between fecundity on the one hand and mortality on the other, with a variety of factors relating to the population size at particular times of the year and the availability of food, particularly in the form of beech mast. She designated eight age classes in the population, the first age class relating to birds in their first year, that is from egg to breeding adult, and the second to the eighth age class representing year classes of birds in the breeding population. Figure 31 shows a flowchart indicating the mechanisms that she built into her model.

At the start of the model, as with most population models, certain critical information must be read in: in this case the date on which the first egg was produced, an indication of the quality and size of the beech mast crop and an initial age distribution. The remainder of the model comprises a large time loop, one pass through the time loop representing one year's population dynamics. Within this time loop, fecundity and mortality events are divided into a temporal sequence and the sequence of calculations summarized in table 25 is carried out.

The first operation in the model is to sum across age classes to give a breeding population at the beginning of summer. The model then calculates the elements for the Leslie matrix used to represent the fecundity process. This matrix is similar to those seen previously, except that the survival elements on the sub-diagonal are all one as adult survival is assumed to be more or less negligible at this time of year. The first row contains age-specific fecundity terms which have to be calculated anew each pass round the time loop. Table 25 contains the basic equation by which these elements are calculated. Simply, the fecundity terms for a particular age class are calculated from the average number of eggs per adult in that age class, computed from the data available to the modeller, modified by functions relating the average number of eggs per adult to the average date of the first egg-laying in the first instance and, in the second instance, to the number of birds in the breeding population, calculated earlier within the time loop. The first two curves in figure 32

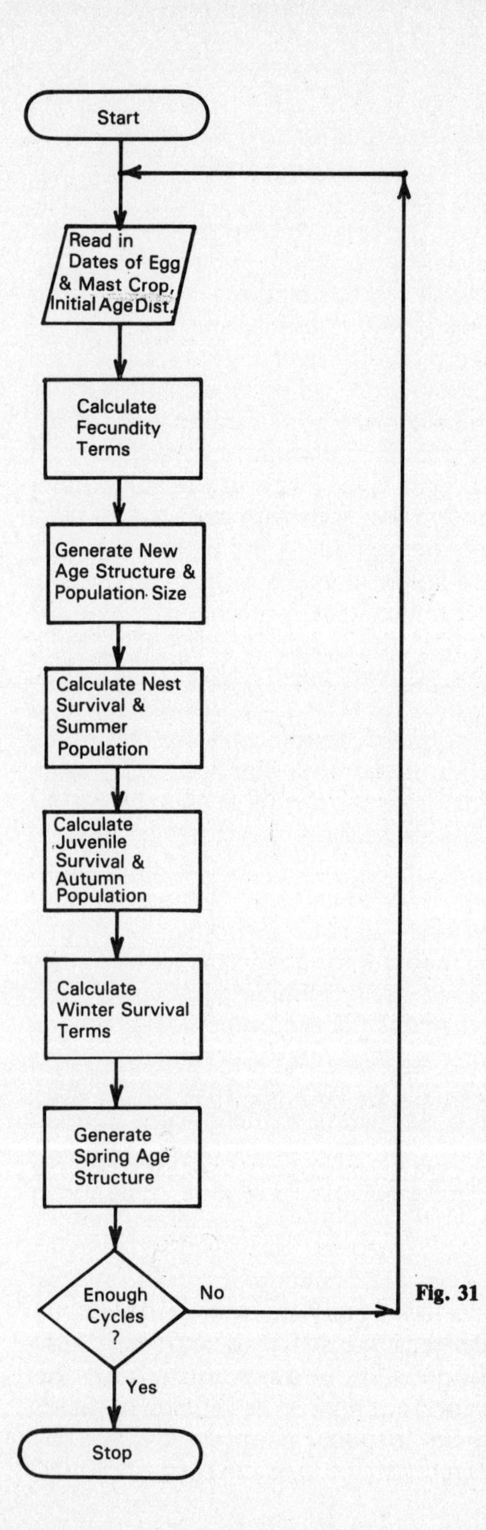

Fig. 31 Flowchart of the model of population dynamics of great tits built by C.L. Pennycuick [after Pennycuick 1969, with permission from *Journal of Theoretical Biology*, Volume 22, © Academic Press Inc. (London) Ltd.]

Table 25. Leslie matrices and other computational devices used in Pennycuick's 1969 model of great tit populations

1. **The fecundity matrix**

$$\begin{bmatrix} F_1 & F_2 & F_3 & F_4 & F_5 & F_6 & F_7 & F_8 \\ 1 & 0 & 0 & 0 & 0 & 0 & 0 & 0 \\ 0 & 1 & 0 & 0 & 0 & 0 & 0 & 0 \\ 0 & 0 & 1 & 0 & 0 & 0 & 0 & 0 \\ 0 & 0 & 0 & 1 & 0 & 0 & 0 & 0 \\ 0 & 0 & 0 & 0 & 1 & 0 & 0 & 0 \\ 0 & 0 & 0 & 0 & 0 & 1 & 0 & 0 \\ 0 & 0 & 0 & 0 & 0 & 0 & 1 & 0 \\ 0 & 0 & 0 & 0 & 0 & 0 & 0 & 1 \end{bmatrix}$$

Where the fecundity terms (F_i) are obtained from

$$F_i = f_i \, \phi(D) \, \phi(N_o)$$

where f_i is the average number of eggs/adults; $\phi(D)$ is a function of the average date of the first egg, D; and $\phi(N_o)$ is a function of the size of the breeding population, N_o.

2. **Nest survival**

$$P_n = p_n \, \phi_2(N_o)$$

where P_n is the proportion of eggs giving rise to fledged young; p_n is mean nest survival; and $\phi_2(N_o)$ is a (second) function of the size of the breeding population.

3. **Juvenile survival**

$$P_j = p_j \, \phi(N_2) \, \phi(BM)$$

where P_j is the proportion of fledglings surviving over their first three months; p_j is the mean juvenile survival; $\phi(N_2)$ is a function of the total summer population; and $\phi(BM)$ is a function of the beech mast crop.

4. **Winter survival matrix**

$$\begin{bmatrix} P_1 & 0 & 0 & 0 & 0 & 0 & 0 & 0 \\ 0 & P_2 & 0 & 0 & 0 & 0 & 0 & 0 \\ 0 & 0 & P_3 & 0 & 0 & 0 & 0 & 0 \\ 0 & 0 & 0 & P_4 & 0 & 0 & 0 & 0 \\ 0 & 0 & 0 & 0 & P_5 & 0 & 0 & 0 \\ 0 & 0 & 0 & 0 & 0 & P_6 & 0 & 0 \\ 0 & 0 & 0 & 0 & 0 & 0 & P_7 & 0 \\ 0 & 0 & 0 & 0 & 0 & 0 & 0 & P_8 \end{bmatrix}$$

where P_i is the probability of overwinter survival of the ith age class obtained from

$$P_i = p_i \, \phi_2(BM) \, \phi(N_3)$$

where p_i is the mean probability of survival; $\phi_2(BM)$ is a (second) function of the beech mast crop; and $\phi(N_3)$ is a function of the number in the population at the start of winter.

indicate the relationships that the Pennycuicks identified in the data and give the equations that were used to represent these functions. Like all the relationships used in the model, the parameter values in the equations of the curves which were fitted to the data were obtained using a least squares method of non-linear regression. Once the fecundity terms have been calculated

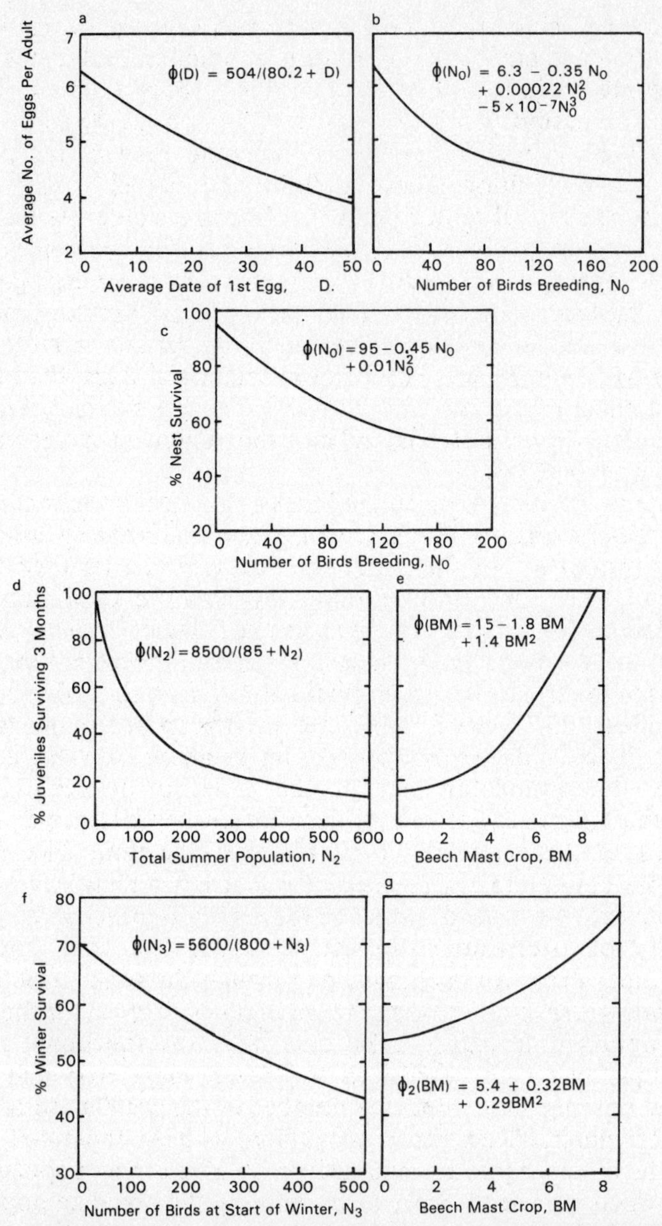

Fig. 32 Relationships built into the Pennycuick model of great tit populations (after Pennycuick 1969): a. egg production versus date of first egg; b. egg production versus number of birds breeding; c. percentage survival of nestlings versus number of birds breeding; d. percentage of juveniles surviving to three months versus size of summer population; e. percentage of juveniles surviving to three months versus size of beech mast crop; f. percentage of overwinter survival versus size of population at onset of winter; g. percentage of overwinter survival versus size of population at onset of winter; g. percentage of overwinter survival versus size of beech mast crop [reproduced with permission from *Journal of Theoretical Biology,* Volume 22, © Academic Press Inc. (London) Ltd.]

for each age class, the model generates a new age distribution using standard matrix multiplication techniques, and then sums the age classes in this "new" population to obtain the post-egg-laying population size. The next two operations in the model are the calculation and imposition of nest and juvenile mortality within the population. These could be represented by matrices, but, as the survival terms involved relate to a single age class in each instance, this would be unnecessarily complicating. Instead, nest survival — that is, the proportion of eggs giving rise to fledged young — is computed by simple multiplication using the functional relationships set out under 2 in table 25. The proportion of eggs surviving is calculated from the mean nest survival and a function related to the size of the breeding population. The functional relationship involved is shown in graph c of figure 32.

Juvenile survival is calculated in similar fashion except that there are two functional relationships involved in the calculation of the proportion of the fledglings that survive over their first three months. The first of these involves the total summer population, that is, the sum of all the age classes in the population after nest mortality has been calculated; the second function relates to the size of the beech mast crop, that is, the food supply available to the juveniles. After the imposition of this mortality on the juvenile age class, the age classes are again summed to obtain the population size at the onset of winter. Overwintering mortality affects all age groups and hence is best applied in the model as a matrix multiplication. The Pennycuicks computed a winter survival matrix which is given as 4 in table 25, where the diagonal of the matrix represents the probability of overwinter survival for each age class and is calculated afresh on each passage through the time loop. This calculation draws on the mean probability of survival for the age class obtained from the basic data and two functional relationships, one with respect to the beech mast crop and the second relating survival to the number in the population at the start of winter. When winter mortality has been computed, the age classes can again be summed up to give a new population size, which represents those birds entering the breeding population the following year. The calculations within the time loop are then complete and the calculations for a second season can be entered into if required.

The additional complexity and biological realism inherent in the Pennycuick model permit more investigation and experimentation to be carried out with the model. The functional forms can be changed, particular processes can be eliminated from particular runs, and the effects on the population dynamics of the species observed. In Pennycuick (1969), the model was used to investigate the relative importance of density dependent and density independent factors in the regulation of natural populations of great tits. We are not concerned here with the findings in this particular area, although the interested reader will find that the paper rewards attention. The important thing is that the combination of numerical simulation techniques, which people like Hardman and others have used, together with some of the more mathematically elegant features present in the Leslie matrix approach, was used to produce a model which was in some ways more satisfactory than either of the two approaches would have been in isolation.

Leslie matrices are closely related to a broader class of modelling techniques utilizing the theory of **Markov chains.** Basically any process which can be represented by an equation of the same type as the basic transition equation:

$$\mathbf{s}_{t+\Delta t} = \mathbf{T}\mathbf{s}_t \quad \text{(see chapter 2)}$$

is a Markov process, so long as a few mathematical constraints on the values within **T** are observed. Hillier and Lieberman (1967) provide a simple introduction to the mathematics and some applications of such processes. In ecology, a number of phenomena have been described as Markov chains in addition to population processes. Usher (1972) describes ways in which energy and nutrient cycling in ecosystems may be represented as Markov chains; Horn (1975) and Shugart and Noble (1981) among others describe plant succession, and Thompson and Vertinsky (1975) analyze the foraging behaviour of birds in this way. Other applications are legion and on the increase.

It follows from comments in chapter 2 that any systems model can be represented using matrix techniques. Virtually all models follow the changes in state of the components within a particular system through time, and hence a basic transition equation is implied. In practice, the size of the matrices and vectors which are implicit in such models may make their direct use impractical (see the aphid models in 6.4 or the ecosystem models in 8.2 for instance). The mathematical condensation involved in reducing ecological processes to matrix form may also make the construction and understanding of such models more difficult for those ecologists who remain, primarily, practical biologists.

6.4 Aphids from Two Continents: Life System Models of Animal Populations

The models examined in this chapter so far have been attempts to mimic relatively simple systems. That is, the systems ecologist in each instance has taken a relatively small set of components from the larger system under study, and has used this simple set to define the area of interest, which he has then proceeded to model. Generally speaking, the primary biological component has been the subject species, with a varying number of components from the surrounding environment such as temperature, levels of food availability and the like. In no case examined so far have any predators been involved; the dynamics that have been modelled have been largely the result of interactions between the biology of the subject species and the physical characteristics of its environment.

Early in the history of systems ecology, workers found that many of the systems they wished to study could not be so simplified: more complete sets of components had to be taken into account if there was to be any hope of producing a useful and adequately realistic representation of the dynamics of particular populations. Frequently, this occurred where studies concerned species of economic importance, where simplification was not possible for pragmatic reasons. A particularly productive approach in the search for more complex and realistic models was initiated by a group of entomologists working in Australia and Canada, who based their models on the **life system**. In chapter 1 this term was defined as the subject species in a population study, plus those components of the environment which interact strongly with that species. Figure 33, taken from Clark, Kitching and Geier (1979), summarizes the constitution and functioning of such systems. The models produced from this school are a development of the book-keeping type encountered in the work of Hardman, and are concerned to follow the changes in abundance and age structure of subject populations through time. The models are characterized, however, by a greater complexity and the facility to handle change in the biological characteristics of the species, which may occur through the course of a season, in response to a variety of environmental factors. For this reason they have been called by N. Gilbert and his co-workers **variable life table models.** This will bring to mind the last approach encountered in our consideration of models based on the Leslie matrix configuration, where the mortality and fecundity factors in the transition

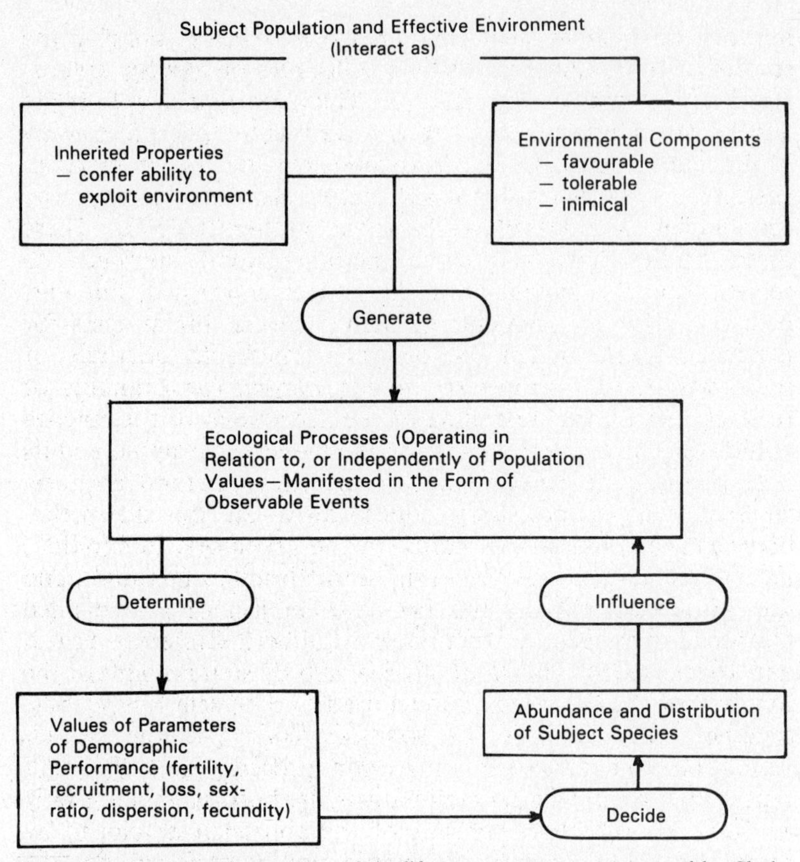

Fig. 33 A diagrammatic summary of the life system concept as presented by Clark, Kitching and Geier (1979)

matrix were variables responding in some predictable way to various environmental factors. The variable life table approach, however, has been carried to somewhat greater lengths than were foreshadowed by the work of Pennycuick, and have produced FORTRAN-based models which are more flexible if less elegant than the matrix-based ones.

Probably the greatest success in this approach to the modelling of insect life systems has been that concerned with the population dynamics of aphids. The first notable exercise in this area was carried out by R.D. Hughes and N. Gilbert (Hughes and Gilbert 1968; Gilbert and Hughes 1971), who modelled the dynamics of the cabbage aphid, *Brevicoryne brassicae*, basing their activities on extensive studies made earlier by Hughes (see for example Hughes 1963). Their efforts were followed up in a number of systems centering on other species, notably the studies of A.P. Gutierrez and his colleagues on cowpea aphids, *Aphis craccivora*, in Australia (see Gutierrez *et al.* 1971, 1974a and 1974b). Gutierrez was also involved with Gilbert in a study of the thimbleberry aphid, *Masonaphis maxima* — notable as an exercise in ecological analysis rather than being primarily economically motivated (Gilbert and Gutierrez 1973). These three studies have the same methodological approach to representing the dynamics of the species concerned, but they pick out different segments of their subjects' life systems for particular attention. This section is concerned with the studies of the cowpea and thimbleberry aphids, chosen for a number of reasons: firstly, the modelling of the life system of the cowpea aphid concentrated on the interaction between the aphid and its host plant, largely neglecting the effects of natural enemies, whereas the second study concentrated on the interaction between the aphid and its parasites and predators. In addition, both examples are well written-up in the primary literature, and the thimbleberry aphid has been the subject of an extended treatment in the synthesizing book by Gilbert, Gutierrez, Frazer and Jones (1976). This book gives a step by step account of the building of the life system model used to represent this species, together with examples of FORTRAN code and the various validation and parameter estimation procedures. I shall not repeat their treatment in detail but will use both examples jointly to draw out the characteristics of the approach developed by these and other workers. It should not be thought that the approach has been restricted, however, to the study of aphids; work is in progress on a number of other species including armyworms (*Heliothis* species), bushflies (*Musca vetustissima*), rabbits and others.

Before going into details of the models, we must consider briefly the basic biology of aphids, to which an admirable introduction is provided by Dixon in a booklet published in 1973. Aphids are plant-sucking bugs which are usually associated with one or a small group of species of plants. They show a wide variety in their life cycles due to their ability to produce both sexually reproductive and parthenogenetic individuals. In addition, each of these forms may be either winged or wingless. The sequence of such morphs is a response to the environmental conditions pertaining to that particular location. For this reason, some forms which have complex life histories with both sexual and asexual reproduction in one part of the world show a much simplified life history when displaced elsewhere. Colonies of aphids on the growing parts of plants may be established in one of two ways: they may be the result of the emergence of nymphs from overwintering eggs, or they may result from the arrival on the plant of a winged individual whose subsequent parthenogenetic reproduction results in the formation of the new colony. After several generations of parthenogenetic reproduction, the colony may build up to very high numbers and winged individuals are produced in response to these high densities. These leave the colony and can establish new colonies on surrounding plants or even in distant habitats. In those individuals where sexual reproduction is part of the life history,

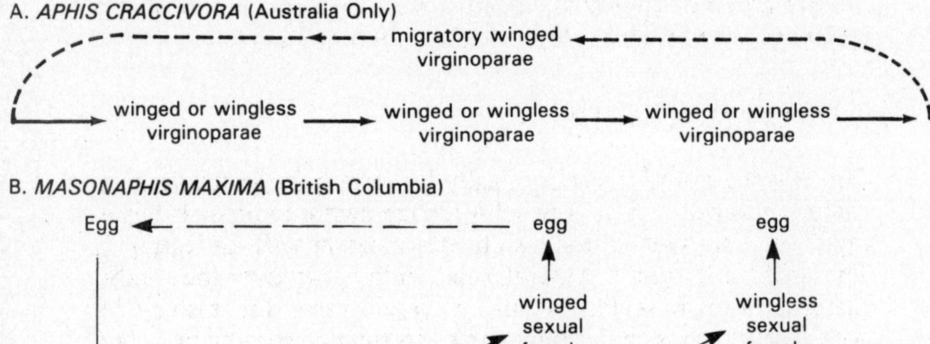

Glossary of terms used in figures and related text.

Alates:	winged morphs
Apterae:	wingless morphs
Fundatrix:	morph responsible for colony foundation at the season's onset
Virginoparae:	parthenogenetically reproducing female morphs
Gynoparae:	female morphs responsible for the production of sexually reproducing offspring
Parthenogens:	females responsible for founding new sub-colonies by asexual reproduction

Fig. 34 Summary of the life cycles of A. the cowpea aphid (*Aphis craccivora*) in Australia and B. the thimbleberry aphid (*Masonaphis maxima*) in British Columbia (after Gilbert and Gutierrez 1973)

the sexual morphs arise towards the end of the season and are responsible for producing overwintering eggs. Most species of aphids are restricted to a single host plant during their life history, but some must alternate between one host and another in order to complete the annual cycle. The migration of the winged morphs of many aphids over long distances has been the subject of much study, a summarizing account of which is given in Johnson (1969). The facility of many species to cover long distances as part of the wind-borne aerial "plankton", accounts for the world-wide ubiquitousness of the group, and it is for this reason that they are of considerable economic importance. The life histories of the two species to be considered in detail shortly, are presented in figure 34, together with an explanation of the technical terms used to describe the different stages.

6.4.1 The Cowpea Aphid

The cowpea aphid is economically important in Australia as a vector of various viruses which damage pasture crops. It has a range of host plants but all can be described broadly as legumes. The aphid is exotic in Australia but widespread over the south-eastern pastoral region. It has a very simple life history in Australia which includes only parthenogenetic reproduction, the sexual stages recorded from its native south-eastern Europe not having been observed in Australia. This species and its associated virus disease, subterranean clover stunt virus (SCSV), were the subject of an intensive study by A.P. Gutierrez and his colleagues based in Canberra. Their findings are summarized in a series of papers (Gutierrez, Morgan and Havenstein 1971, Gutierrez, Havenstein, Nix and Moore 1974a, and Gutierrez, Nix, Havenstein and Moore 1974b). The 1974a paper describes a simulation model of the population dynamics of the aphid and its food plant in pasture regions of south-eastern Australia.

The model built by these authors is illustrated by the flowcharts presented in figures 35 and 36. The key feature of their model was that it was basically a representation of the dynamics of the food plant (fig. 35), with a submodel representing the aphid populations forming a small part of this larger plant model (fig. 36).

Before examining the model in detail, it is necessary to look at the way Gutierrez and his colleagues handled the time problem in their model. As in other models of insect dynamics, the developmental and reproduction processes in the cowpea aphid

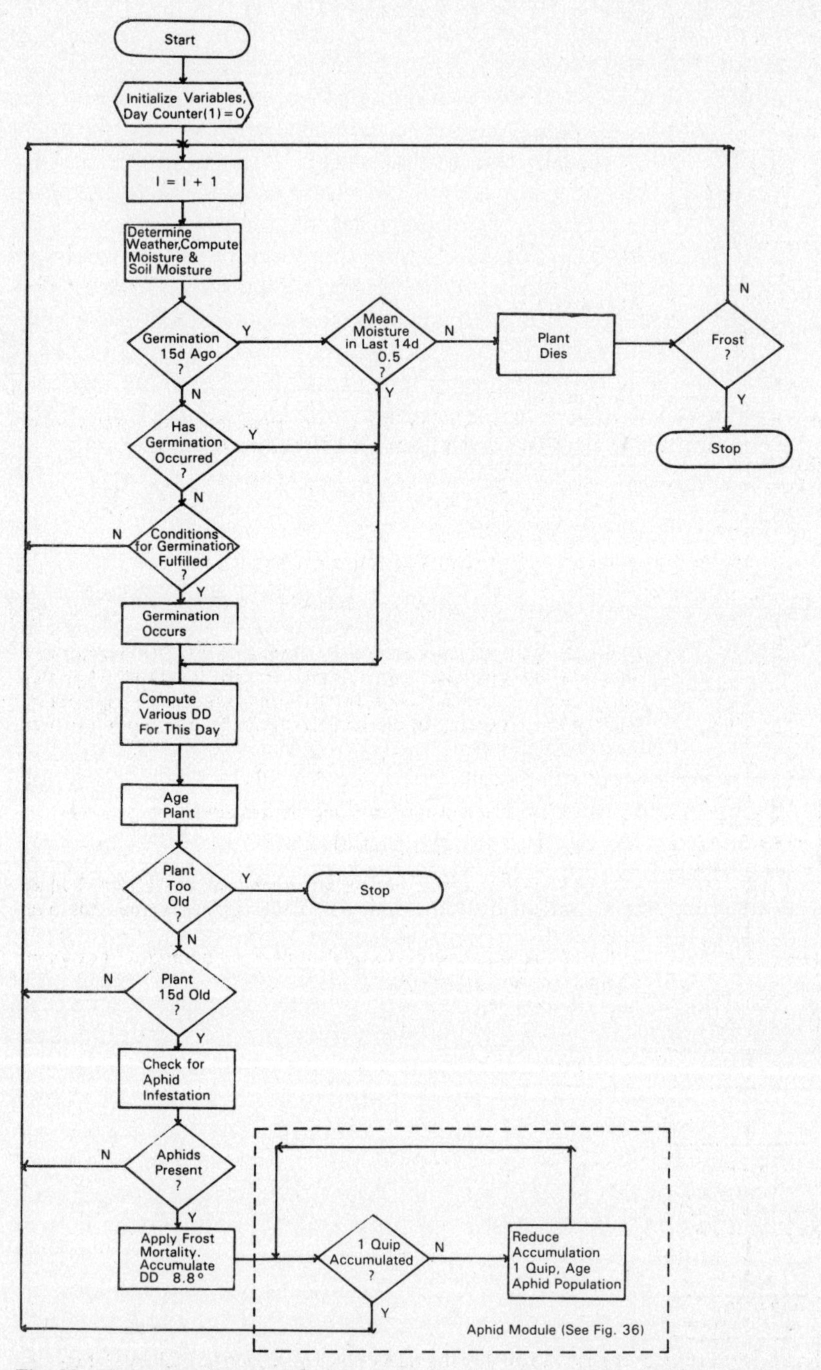

Fig. 35 Flowchart of the model of the dynamics of temperate legumes built by Gutierrez *et al.* (1974a)

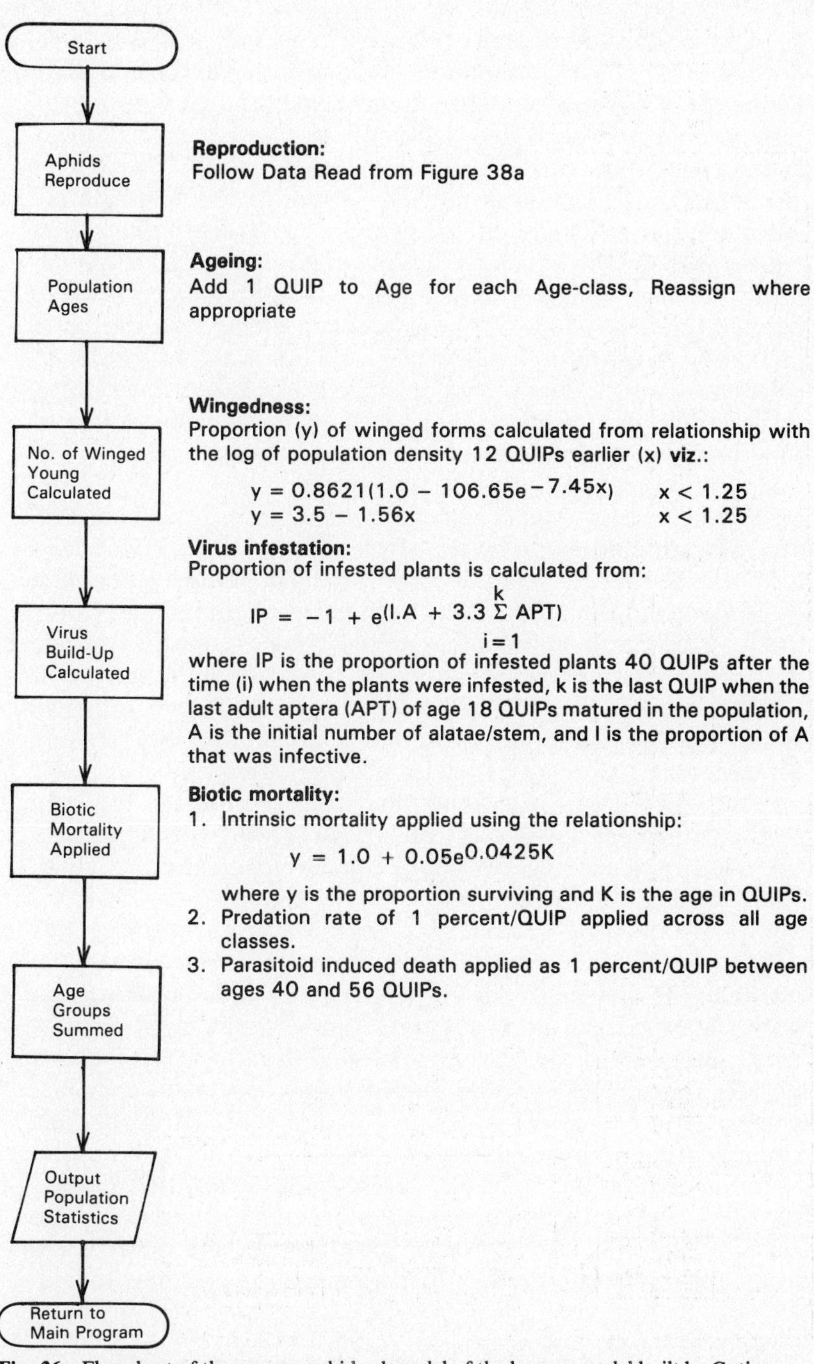

Fig. 36 Flowchart of the cowpea aphid submodel of the legume model built by Gutierrez *et al.* (1974a)

are temperature-dependent, hence all time scales used in connection with their population dynamics have to be made temperature-dependent. This is achieved by calculating the durations of particular age classes of the insect in terms of day degrees (D°), as with Hardman's flour beetles. These were calculated in the standard manner, by rearing the aphids in the laboratory at a variety of temperatures and determining that temperature below which no significant amount of development occurred. For the cowpea aphid this developmental zero temperature was 8.3°C.* Developmental changes in the plant are also temperature-dependent, and it was necessary to calculate a separate set of day degrees for the plant. Experimental methods very similar to those outlined for the aphids showed that the plants concerned had a developmental zero of 5.5°C. In addition to these two basic biological time scales, Gutierrez and his colleagues also used the calendar time in days when considering the germination processes of the plant. They incorporated into their model relationships between temperature and moisture conditions and the likelihood of germination in the plant, using data on the biology of the legumes concerned which had been reported elsewhere in the literature by a variety of earlier authors. One last time scale was required in applying frost-induced mortality to the aphid populations. Gutierrez and his co-workers were able to calculate a relationship between the percentage of aphids surviving in any time period and the ratio of the number of day degrees calculated **below** freezing point to the number of day degrees **above** the developmental threshold. They adopted the figure of 2.2°C as the threshold for the freezing point in their calculations of these frost-mortality day degrees, because the grass temperature was always a few degrees less than the Stevenson Screen temperatures which were used as basic data for their work. Figure 37 shows the relationship that they used between the survivorship and this ratio of the two day degrees scales. The three basic temperature dependant scales are computed in the model from information collected in the field over the time period of interest.

In the calculation of the day degree sums for any day, Gutierrez *et al.* used a technique which is becoming standard in this sort of work: they fitted a sine curve between the shade maximum and minimum temperatures obtained for each day, and in-

---

*The reader referring back to the original work will note that Gutierrez *et al.*, used degress Fahrenheit in their calculations. I have converted these to degrees Centigrade for the sake of consistency.

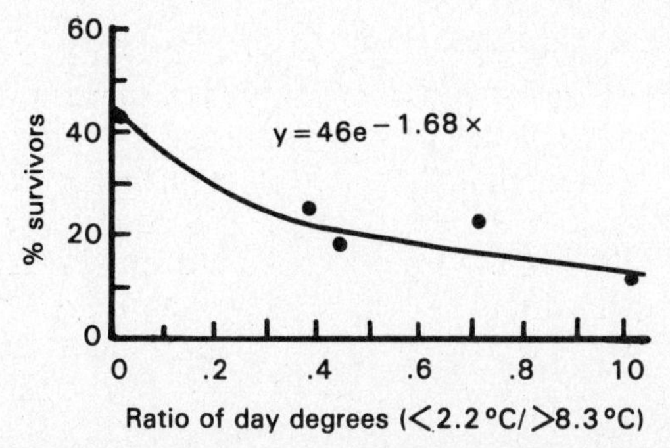

Fig. 37 The empirical relationship between percentage survival of aphids on the ratio of day degrees below 2.2°C to those above 8.3°C, used by Gutierrez *et al.* (1974a) to compute frost induced mortality in cowpea aphids

tegrated above the particular developmental threshold to obtain the plant and aphid day degree sums, and below the freezing point threshold in order to obtain the day degree sums relating to frost mortality. It has been shown in a number of instances that this sort of fitting technique, based on the relatively small amount of information inherent in daily maximum and minimum temperatures, is sufficiently accurate to provide adequate predictions in developmental models of this kind. It is possible to obtain much more elaborate and, theoretically, more accurate temperature scales for a particular area by using various continuous or hourly recording devices, but the amount of data handling is frequently not justified by any substantial increase in predictive accuracy.

There is a general moral to be drawn from this situation: frequently the amount of data needed to provide adequate input to a simulation model is much less than may be available or than can be gathered under certain circumstances. Some sort of rule of parsimony is certainly appropriate here, as the sum total of input required for a model is usually substantial and a certain efficiency in the collection of data vis-a-vis any one of these inputs is highly desirable.

In addition to these practical considerations, Gutierrez and his colleagues made a decision about the **step size** to be used in the part of their model relating to the dynamics of the aphid populations. They argued that from a practical point of view it was far better to work in units of more than one, or any other arbitrarily small number of day degrees, in ageing the population in any particular pass through the aphid submodel. Accordingly, they worked in terms of **quarter instar periods** or **QUIPS**, each of which was defined as a quarter of the total time taken for the development of the first instar aphid nymph. They then rescaled all their other developmental times into numbers of QUIPS and only carried out an update on the aphid population structure in their model every QUIP. In fact, for the cowpea aphid one QUIP was 4.7 day degrees over the developmental threshold temperature of 8.3°C. Other biological information on the cowpea aphid was calculated, from experimental observations or from the literature, onto a time scale based on QUIPS. In particular, the reproductive rate per unit time and the survivorship curve were reworked in this fashion and figure 38 shows the relationships used for each of these parameters. This standardization in the time step for the ageing process is in contrast to the approach of Hardman, for example, who used a **calendar** time step involving, accordingly,

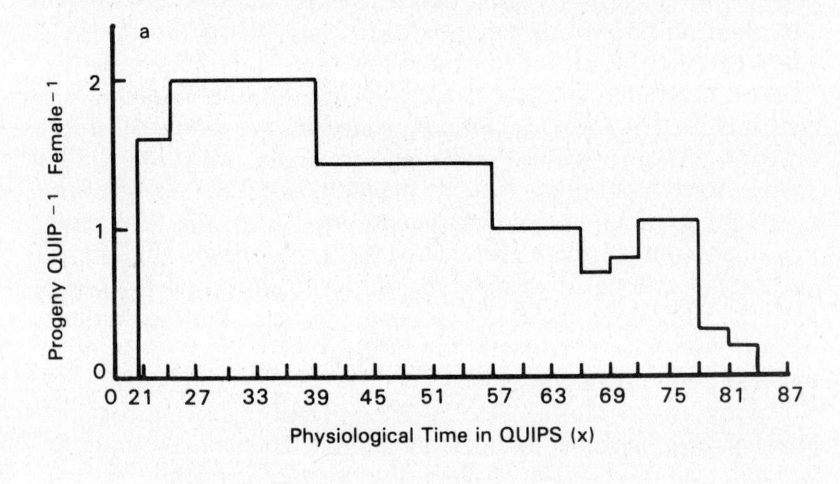

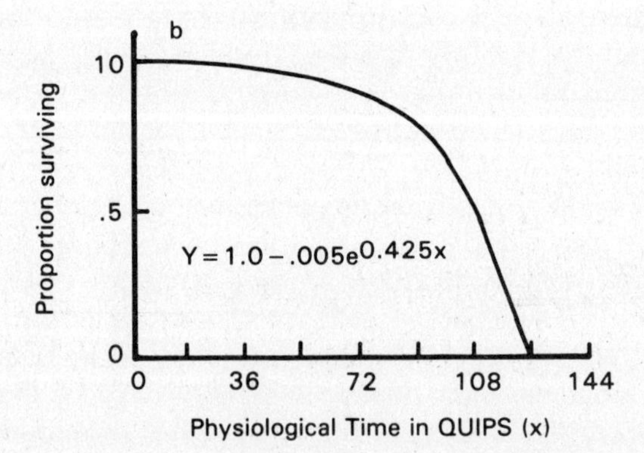

Fig. 38 a. reproduction and b. survivorship curves for cowpea aphids, with time scales expressed in QUIPS (after Gutierrez *et al.* 1974a)

variable numbers of day degrees on each passage through the time loop of his model.

We can now examine the model of the dynamics of the legume illustrated in the flow diagram, in figure 35. Essentially the model follows the progress of a particular crop of the legume and is based on the responses of individual plants to the various biotic and abiotic factors which affect their germination, survival and growth, including the depredations of the aphids which are associated with the plants.

The model is based on a large time loop operating on a day to day basis. The first operation is to initialize the day counter to zero. Then the major time loop is entered. The first segment of computations relates to the germination of the plant. Whether or not a plant germinates in a particular period of time depends upon the combination of rainfall and temperature which has occurred in the preceding few days. If conditions suitable for germination have occurred at some time in the past, but the plant then encounters a sequence of rainfall conditions which lead to excessive moisture stress, the plant will die back. Subsequently, as long as no frost occurs, the plant can regerminate and grow a new set of foliage. If, however, the poor moisture conditions are followed by or are concurrent with a frost then the plant dies and, in the model, this is represented by the termination of the particular run of the simulation. The actual mechanics of the germination process are well illustrated in the flowchart, and the details of the conditions necessary for successful germination are available in the paper by Gutierrez *et al.* (1974a).

Leaving the germination routine of the model, the flow of computation next encounters the code which calculates the various day degree sums. The three different sums are calculated using fitting techniques, and provide the basic information on which the rest of the model depends at several points.

The next sequence of computations is called the **plant ageing routine**. This adds the number of plant day degrees encountered in a particular day to the running total, maintained elsewhere, of the age of the plant. If this sum is greater than the observed maximum survival time of the plant, the run is terminated; if not, the plant is tested to see whether it is more or less than fifteen days old. Plants less than fifteen days old are not available for infestation by aphids, so the flow through the model returns to the beginning of the time loop. Older plants may be infested by aphids, in which case the flow of computation then enters the sequence of processes which includes the model of aphid dynamics. If there has been an aphid infestation, either initiated during this pass through the time loop or resulting from an earlier event, then the overall population of aphids is reduced, if necessary by applying the frost mortality already alluded to and employing the relationship given in figure 37. In addition, the number of aphid day degrees accumulating in this pass through the time loop is added to the cumulative total of such day degrees previously calculated. If the accumulated total so calculated is more than 4.7, that is, if it is more than one QUIP, then the aphid submodel is called, and the dynamics of the aphid population during that particular QUIP are followed in a fashion which will be described shortly. Following this application of the aphid submodel, the accumulated total of aphid day degrees is reduced by one QUIP, which portion of time has already been "used up". The flow of computations returns to the previous test at which the accumulated total is again tested and, if there remains another QUIP's worth of day degrees, then the model relating to aphid dynamics is called yet again. This looping device takes care of the situation where more than 9.4 aphid day degrees (two QUIPS) accrue in one day, which is possible under certain circumstances.

The aphid submodel illustrated by the flowchart in figure 36 updates the age structure and the number of aphids associated with the plant which occur within the period of one QUIP. In addition, the submodel predicts the proportion of plants infected by the virus of which the aphid is the vector. Within this submodel, the first procedure is the reproduction process, which is achieved by calculations based upon the data shown in figure 38a. This relationship between fecundity and physiological time is read into the initializing part of the model as a data statement: that is, the actual figures collected from experimental work are used in the model, and the number of progeny produced by the aphid population is calculated on the basis of the age-specific reproductive rates contained in them. The whole population of aphids is then aged using methods similar to those encountered before, whereby the elapsed time period, one QUIP, is added to the age for each age class, and organisms are reassigned to older age classes where this process causes them to move out of one class into the next. The next calculation carried out within the submodel is the determination of the proportion of winged forms appearing in the fourth nymphal instar. Gutierrez and his colleagues were able to identify a relationship between this proportion and the population density of all age classes together, twelve QUIPs earlier — that is, the population density at the time at which the current fourth instar nymphs first entered the population as newborn individuals. The determination of this proportion is of particular importance, as it is the winged forms which provide the propagules which initiate infestation in adjacent plots of the pasture. The relationships used are complex and non-linear and are given in figure 36, adjacent to the appropriate box in the flowchart. Following these procedures the submodel calculated the state of virus build-up in the plants. We are not concerned here with the dynamics involved, suffice it to say that a complex relationship was identified between the proportion of infected plants and the age structure and size of the aphid population, of which a known proportion was infective. The reader is referred to Gutierrez *et al.* (1974a) for further details. The submodel then returns to an ecological process within the aphid system and applies the effects of biological mortality agencies known to act upon the population. These take three forms and each is applied to the population separately. The first is the intrinsic mortality applied from the relationship shown in figure 38b, in which the survivorship curve of the individuals obtained from laboratory studies is presented. The formula involved is a simple negative exponential relationship, and is given with the figure. Following the application of this intrinsic mortality, a mortality due to predation, of 1 per cent per QUIP, is applied across all age classes. This figure is based upon earlier observations made by Gutierrez's group (1971), which showed that predators had very little effect on cowpea aphid populations. In addition, some parasitoids do induce death in a small proportion of individuals between the late second instar and fourth instar stages, and a mortality is applied to these age groups in the model, again at the rate of an additional 1 per cent per QUIP. The remainder of the aphid submodel is simply a book-keeping device, whereby the numbers within each age group are summed and the population vector is written out. Following these calculations, control returns to the primary plant model and the test referred to earlier is made to see whether the next stage of operation should be another pass through the aphid submodel or a return to the beginning of the primary time loop within the overall plant model.

Gutierrez and his colleagues validated the final version of their model against field results obtained from three separate sites in New South Wales. They found an acceptably close correspondence between their predictions and the phenology of all

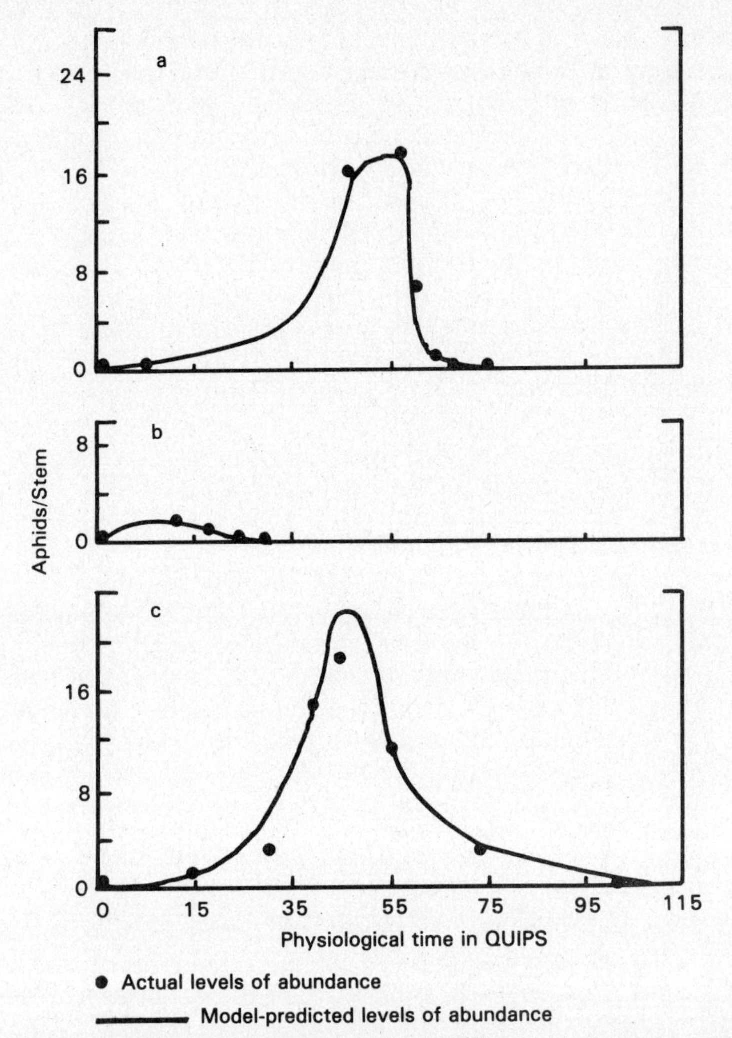

Fig. 39 Comparision of actual and model-predicted levels of abundance of cowpea aphids at three separate sites in New South Wales, (a) Bathurst, (b) Orange and (c) Trangie (after Gutierrez *et al.* 1974)

three aphid populations when the environmental data from those three sites was used to drive the model. Figure 39 shows a comparison between the predicted and observed population trends in the three sites.

The model was used to simulate the effects of varying fecundity rates, the infestation rates, and the intensity of various mortality agencies, particularly predation and parasitism. From these considerations the authors were able to draw conclusions about the strategies employed by the insect in order to maximize its success, and also to make recommendations with respect to the potential for biological control of the aphid in Australia by the introduction of its natural enemies from other parts of the world.

6.4.2 The Thimbleberry Aphid

The thimbleberry aphid is an inhabitant of the Pacific north-west of North America where its food plant, *Rubus parviflorus*, is a common component of the shrub layer of coniferous forests. The aphid is restricted to this host plant and is of no economic importance as the plant, although producing edible fruits, is not grown domestically. The aphids restrict their activities to the less woody growing parts of the plant and can be sampled by removing these "terminals" selectively from the whole mass of canes constituting the growth form of the plant. The aphid has a complex life history illustrated in figure 34b and develops both parthenogenetic and sexually reproductive forms which can be either winged or wingless. These different morphs also differ in colour and can be identified readily in this fashion. Colonies develop at the beginning of the season from overwintering eggs, and additional colonies may arise throughout the season as a result of the migration of the winged parthenogens (aphids capable of founding new colonies by asexual reproduction) leaving the mother colony and moving to adjacent or distant uninfested plants. Gilbert and Gutierrez (1973) observed five discrete generations between March and July in the Vancouver region and, consequently, eggs were produced in each of these generations.

These authors carried out a population study of the species, recording numbers and age structures from a sample of growing tips of the plant throughout one season. They also recorded rates of predation and parasitism involving predatory syrphid larvae and parasitic Hymenoptera, belonging to the wasp species *Aphidius rubifolii*. Of these natural enemies they concentrated on the parasitoid, and carried out a parallel study on the biology and ecology of this species. The parasitoid larvae develop and pupate within the infested aphids and may or may not enter a period of suspended animation or **diapause** to overcome unfavourable periods. They recorded the proportion of diapausing parasitoids throughout the season and also the rate of attack by hyperparasites, that is parasitoids which attack the parasitoids already contained within aphid "mummies". The authors synthesized all this information, together with laboratory-based information on rates of development of both the host aphid and parasitic wasp by building a simulation model of the type we have already encountered. Their 1973 paper and the book by Gilbert, Gutierrez, Frazer and Jones (1976) give a full account of this modelling exercise.

Figure 40 is a flowchart summarizing the steps that Gilbert and Gutierrez built into their model. Once again the QUIP was used as the basic time unit, and the main loop of the model shown on the left of the figure is similar in many respects to the aphid module built by Gutierrez and his colleagues into their representation of the dynamics of cowpea aphids. The model of the thimbleberry aphid takes little account of the state or dynamics of the food plant. This is probably a justifiable simplification in this case as the dynamics of the aphid through one season occur at a time of superabundant and high quality food availability. The establishment and build-up of the colony in spring occurs during the period of most rapid growth of shoots of the thimbleberry, and the aphids complete their active period of reproduction and dispersal about midsummer, well before the end of the growing season for the plant. Thus, although the thimbleberry plant does have responses to the environment and dynamic properties which are not independent of the driving variables involved in aphid reproduction, Gilbert and Gutierrez made the decision to ignore these factors because their effect upon the aphids was likely to be very small on the time scale and at the level of resolution at which they chose to work. Unlike the model of the cowpea aphid, however, these authors chose to add to their basic model of aphid dynamics, detailed versions of the effects of parasitism and predation within the survival segment of the model of aphid populations.

The flowchart shows that the aphid model simulated three major processes: **reproduction, ageing** and **survival**. Each of these occurred within the time loop representing the passage of real time throughout one season. In the **reproductive** process, the number of offspring appearing in each time interval was computed and the changing morph frequencies calculated over the same interval. Gilbert *et al.* (1976) give details of both the biology of reproduction and morph determination in this species of aphid, and the mechanisms by which these biological characteristics were incorporated in the model. Table 26 shows the procedures involved. Following this reproductive process, the **ageing** process of the aphids is simulated, adding one QUIP to the age of each, removing seniles and recomputing morph abundances following the ageing process.

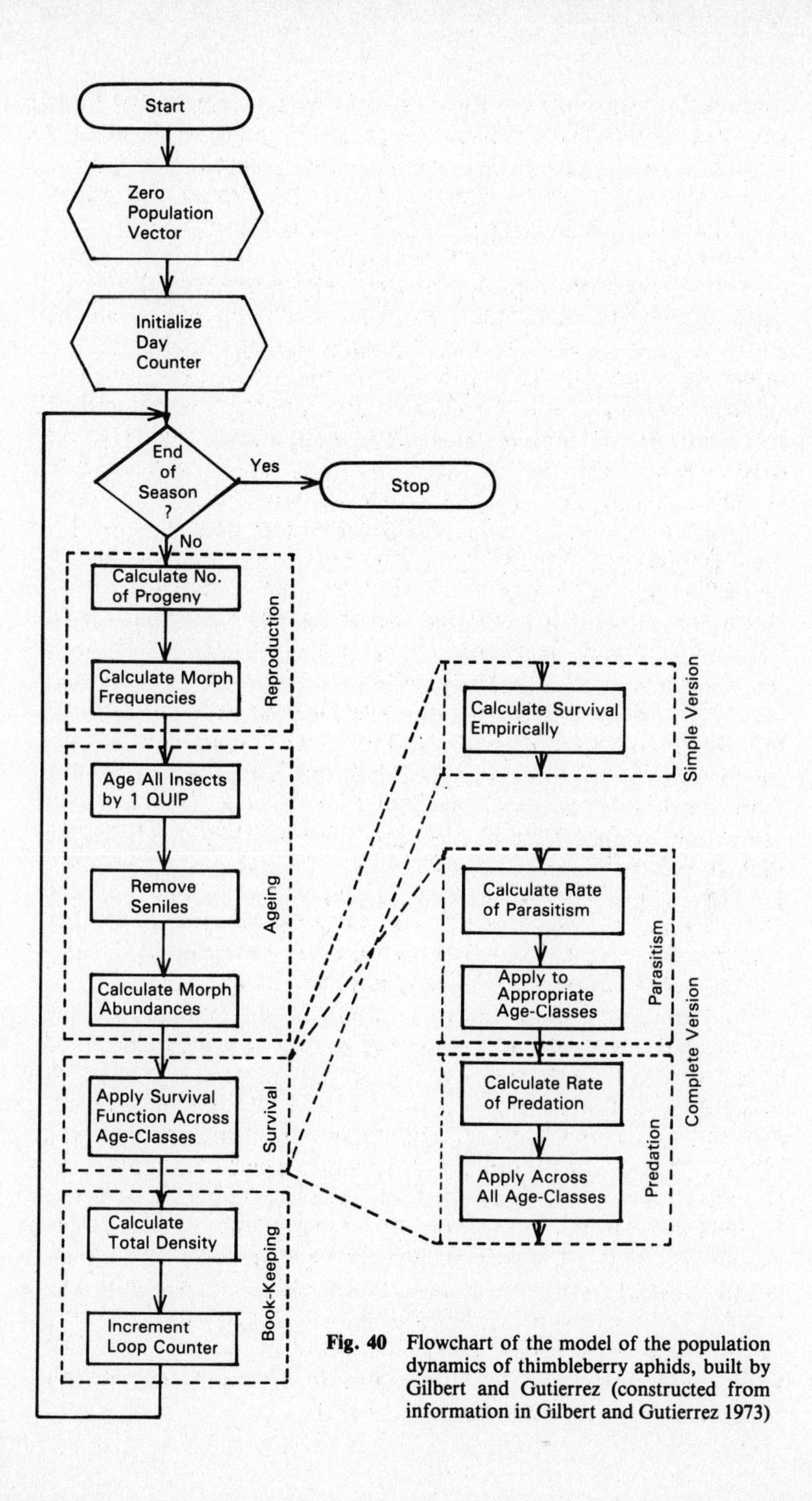

Fig. 40 Flowchart of the model of the population dynamics of thimbleberry aphids, built by Gilbert and Gutierrez (constructed from information in Gilbert and Gutierrez 1973)

Table 26. The reproductive process in the thimbleberry aphid model, first version (after Gilbert *et al.* 1976)

1. Determine the fecundity level/parthenogen for the particular time step.
   Determined by: (a) Time of season.
   (b) Current population density.
   Based upon empirically determined relationships.

2. Determine the morph frequency within the predicted offspring/parthenogen.
   (a) Proportion of gynoparae determined from the number of females in the next generation.
   (b) Proportion of males is related to the time of season.
   (c) Proportion of parthenogens is calculated by difference.
   (a) and (b) are formulated using empirical relationships.

3. Calculate absolute numbers of offspring of each type knowing:
   (a) The proportions calculated in (2).
   (b) The age and morph specific numbers of reproductive adults (computed in the model or, initially, read in).
   (c) Current individual fecundity levels calculated in (1).
   (d) The age-specific reproductive patterns (measured empirically).

The next segment of the model of Gilbert and Gutierrez is perhaps the most interesting and the most innovative in comparison with previous models of aphid populations. The **survival** module, shown in its simplest form on the left hand side of figure 40, comprises the application of a simple survival function across all age classes. This is based on an empirical procedure by which mortality is partitioned across all age classes measured in the field, similar in some ways to that used by Hardman (see table 18). This process is represented in table 27. In Gilbert *et al.* (1976) the first version of the model that they

Table 27. The survival routine used in the thimbleberry aphid model, first version (after Gilbert *et al.* 1976)

1. For the first 62 QUIPS of the season mortality is zero and, accordingly, survival is 100 per cent.

2. After the 62nd QUIP, the survival/QUIP/age class is determined from the equation:

$$SURV = (1 - .0001425\,(K\text{-}62)^2 - .000003125\,(K\text{-}62)^3)^{.025}$$

   where SURV is the survival rate and K the counter for the time step (that is, the number of QUIPS elapsed since the beginning of the season). This equation represents an empirical fit to the survival data over the whole of the latter part of the season.

3. Each class of aphids is reduced according to the value of SURV obtained in (2). For example: for the alate stage (ALATE(1))

$$ALATE\ (1) = SURV.ALATE\ (1)$$

   A parallel procedure is applied to the numbers of gynoparae, males, females and apterae.

Table 28. The predation routine used in the Thimbleberry aphid model, second and subsequent versions (after Gilbert *et al.* 1976)

1. Calculate the total number of aphids present at particular time step (= TOT).

2. Calculate the "demand" for aphids (PRED) as follows:
   (a) Calculate the numbers of syrphids in each of the four larval instars at this point in time, given that the numbers of syrphid eggs laid at any time is proportional to TOT at that time and using empirical survival terms for each syrphid instar.
   e.g. The numbers of, say, third instars "now" will be proportional to the sum of the numbers of aphids present on the terminals from 20 to 30 QUIPS ago, reduced by some survival factor.
   (b) Calculate the overall demand, knowing the voracity of the different syrphid instars in terms of numbers of average-sized aphids eaten per QUIP (Instar 1 = 1; Instar 2 = 2; Instar 3 = 6; and Instar 4 = 15).

3. Compute survival using the standard Thompson (1924) formula:

   $$SURV = e^{(-PRED/TOT)}$$

   Symbols as above and in table 27.

4. Apply SURV term across all groups of aphids as in table 27.

describe in detail is of this simpler sort. A second version was constructed which incorporated a predation process. This "improved" version models the effects of syrphid larvae which prey upon the developing aphids, replacing the empirical function with an explicit routine for the action of these predators. This was built into the model by calculating a "demand for prey" per time step generated by the predatory larvae and satisfied by a depletion in the aphid numbers. The authors of the model calculated this demand knowing something about the distribution and numbers of syrphid eggs laid on the terminals where the aphid colonies were and the average voracities of the four instars of the syrphid larvae. These steps are summarized in table 28. The procedures for incorporating this effect into the model of aphid dynamics are an excellent example of the way in which the relatively sparse but critical information available on a particular organism can be incorporated in a sensible and useful way into a model whose prime purpose is to simulate the dynamics of another species. The authors substituted a more complex version for the empirical mortality function present in the first of their models, and then simulated the dynamics of the aphids for the season for which they had field data. This simulation gave reasonable results for the latter part of the season, when predation by syrphid larvae is an important and even dominant agency of mortality.

Gilbert and his colleagues improved further upon this version of the model in the next stage of their work, which incorporated the effects of parasitism into the survival subroutine of the model. The earlier versions of their model ignored the effects of parasitism, retaining as they did parasitized aphids and "mummies" as part of the surviving population of aphids. From laboratory work they calculated the duration of the egg stage and of the larval instars of the parasitoid, together with the rate of hyperparasitism affecting the primary parasitoid throughout the season. Again they assumed random search on the part of the parasitoid and used an empirically determined survival rate for the parasitoids. The actual computational methods included a series of stages involving stepwise improvement of the parameter values used and, in a few cases, simple fitting procedures in order to obtain better correspondence between the output of the model and their field observations. The final routine adopted is described in table 29.

Table 29. The parasitism routine used in the Thimbleberry aphid model, final version (after Gilbert *et al.* 1976)

---

1. Initialize numbers of the different parasitoid stages according to the numbers observed at the beginning of the season.
2. In each pass through the time loop:

(a) Calculate the amount of hyperparasitism according to the empirical, linear equation

PARS = 0.85 - 0.006K

where PARS is survival of parasitoids and K is the time counter as before.

(b) Calculate the proportion of parasitoids entering diapause (DIAP) from the empirical function:

DIAP = 0.02 * (K - 67)

Note this implies that there is no diapause until after the 67th time step.

(c) Calculate potential reproduction of parasites (PAREP) where this is the sum of the numbers in each age class times an empirically determined reproductive rate for each.

(d) Sum the numbers of aphids available for parasitism (X) according to:

X = THREE + FOUR + APTRA (19) + FEM (19)

where THREE is the number of 3rd instars, FOUR the number of 4th instars, APTRA (19) and FEM (19), the numbers of young adult aphids.

(e) Calculate survival of these classes of aphids from parasitism (SP) using the Thompson (1924) formula:

SP = $e^{(-\text{PAREP}/X)}$

Symbols as above.

(f) Compute numbers of surviving adult female parasitoids (i.e. of age 35 QUIPS) (PAR (35)) according to:

PAR (35) = PAR (35) * SR * PARS * (1 - DIAP)

where SR is the observed sex ratio and the other symbols are as above.

(g) The running total of diapausing parasitoids ("mummies" (DITOT) is calculated:

DITOT = DITOT + PAR (35) * DIAP

(h) Numbers of all aphids are adjusted for parasitism.
For example: for the alate stage (ALATE (1))

ALATE (1) = ALATE (1) * SP

a similar procedure is applied to the other age classes.

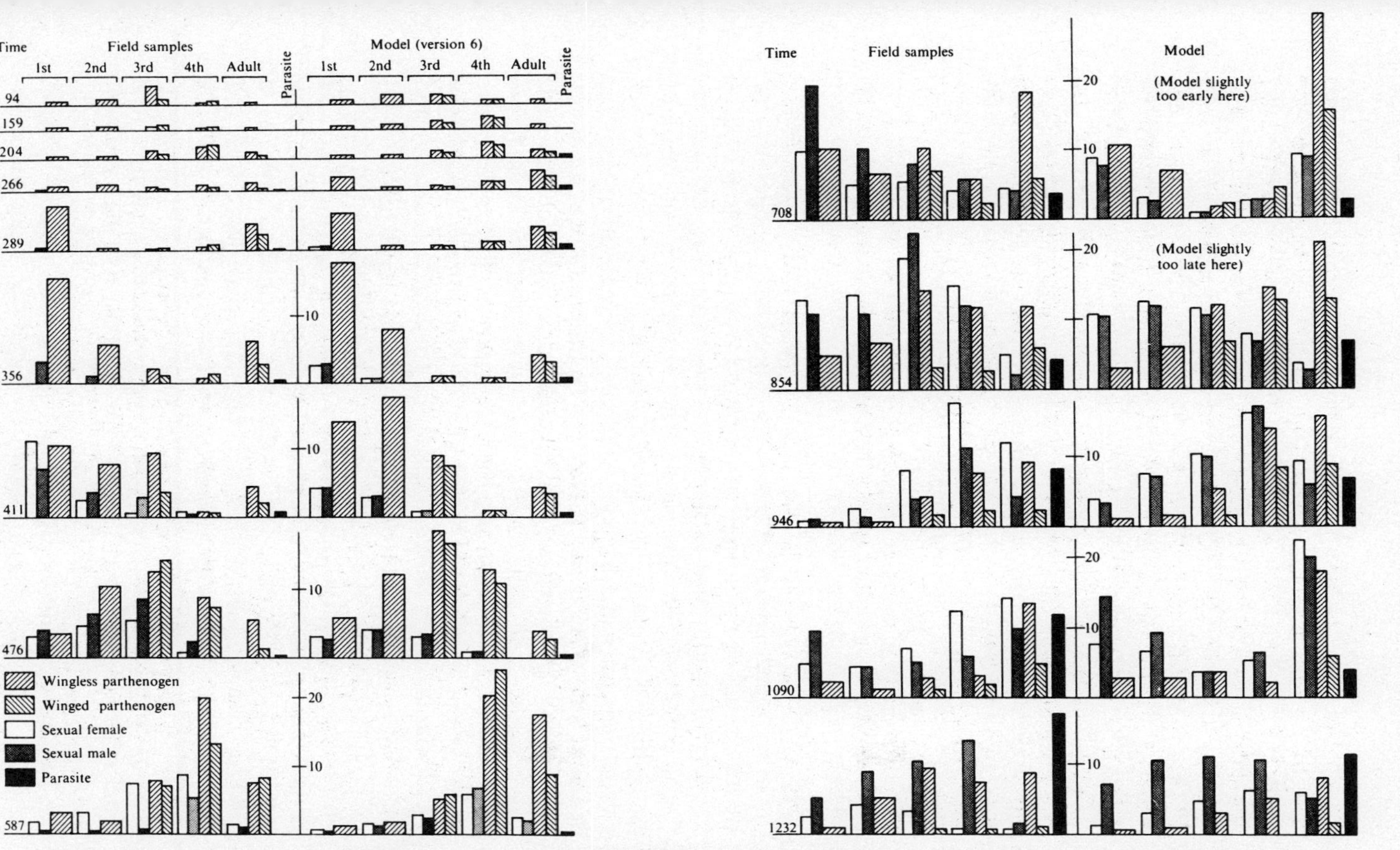

Fig. 41 Comparison of the detailed output of the model of thimbleberry aphid dynamics built by Gilbert and Gutierrez, with field data (after Gilbert *et al.* 1976, *Ecological Relationships,* © W.H. Freeman and Company Ltd). The time scale is in day-degrees elapsed since the beginning of the season

The last major process in the model was a book-keeping one in which the predictions of the model were printed out and a comparision was made between this output and field data for the same season. Figure 41 shows a comparision between the output of the final version of their model on the one hand and field samples for corresponding points in the season on the other. Concerning this comparison and the whole modelling procedure that they adopted, Gilbert and Gutierrez (1973) made the following remark:

> The first version of the program gave answers widely different from the basic data. We made various changes based on pieces of information which we had overlooked, but which were nevertheless present in the data. The model then gave realistic answers. This is the first assurance we get that our understanding is reasonably complete: if something important were missing, we could only reconcile the model to the basic data by large arbitrary interventions. We could, in fact, improve the agreement by various small adjustments, but we prefer not to do so, since such adjustments must be pure guesswork. A model is no more accurate than the biological information which it contains.

Gilbert and Gutierrez used the simulation model of the dynamics of the thimbleberry aphid to investigate the effects of varying the biological parameters involved. They adopted as their criterion for "success" the proportion of sexually reproductive females produced, as these give rise to the fundatrices on which depends the year to year survival of the species. They varied the values for fecundity, age of maturity and morph frequencies, and demonstrated that the observed value in each case is the one which gives the best or closest to the best performance according to their criterion. Figure 42 shows these results in graphical form. They also examined the strategy of the parasitoid in similar fashion by varying its fecundity and concluded that, as a model of the dynamics of the parasitoid, their formulation is weak. Their explorations did indicate, however, that aphid numbers are not very sensitive to the density of the parasitoids which, accordingly, must be assigned a less than central role in the regulation of aphid numbers — a finding which could be of considerable signficance if management by biological control were in question.

Gilbert and Gutierrez concluded their account with a discussion of "fitness" and the implications to be drawn from their simulation experiments about the fitness of their subject population. They proposed the analysis of ecological strategies by parameter manipulation as a way of verifying simulation models, veracity being indicated by close correspondence be-

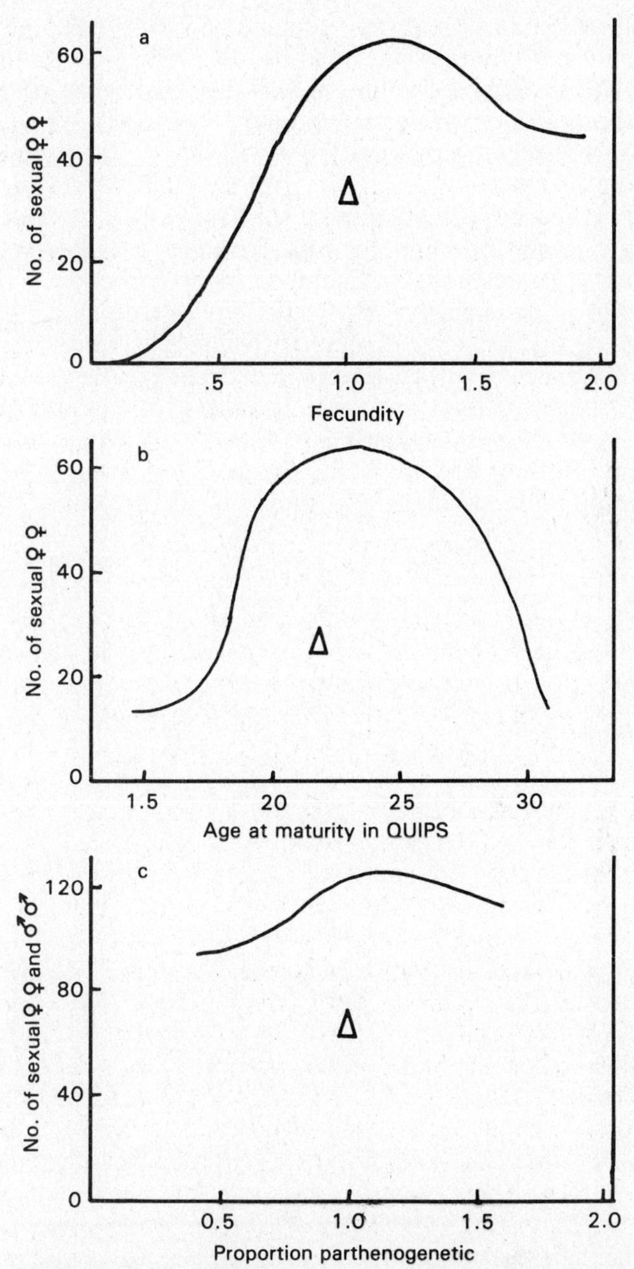

Fig. 42 Predictions of the thimbleberry aphid model of Gilbert and Gutierrez. a. the number of sexual females versus rate of fecundity; b. the number of sexual females versus age of maturity; and c. the number of sexual males and females versus the proportion of parthenogenetic aphids in the population. The arrows indicate the average value for the parameters involved, observed in the field (after Gilbert and Gutierrez, 1973)

tween predicted optimal performance and that indicated using measured parameter values. They conclude: "A simulation model has no intrinsic value. It is useful only when it exposes our ignorance, or answers biological questions. In our experience the greatest difficulty is to formulate the appropriate questions."

The approach by these authors to the process of predation in natural systems underlines the need for a detailed understanding and careful analysis of the general ecological processes of which predation is an example. This need has proved susceptible to analysis using systems ecological techniques and chapter 7 illustrates one very powerful approach to the problems involved. Later work by Gilbert and his colleagues (for example, Frazer and Gilbert 1976; Baumgaertner *et al.* 1981) converges on earlier work by Holling and his students, bridging the gap between population and individual level studies of the predation process.

7

Ecological Processes

Chapter 6 examined a series of models of biological populations, which attempted to simulate the changes in a set of state variables representing particular populations through time. The changes taking place in the value of a particular state variable over a given time span represented the actions and interactions of one or more ecological processes. These processes, such as reproduction, competition, predation, movement and so on, were summarized in the equations of change governing the transition from one point in time to the next for the set of state variables. Frequently, these representations were in abbreviated or condensed form so they would fit into the population level format of the overall simulation. The ecological processes themselves have long taken the interest of research workers, and it is not surprising therefore that, shortly after the introduction of techniques of systems analysis within ecology, these techniques should be taken and applied in a detailed study of one of these processes. Historically the first such study was of predation, the work of C.S. Holling, and his techniques, models and philosophy will be discussed in the second part of this chapter.

Ecologists from Solomon (1949) onwards have recognized a distinction between what has become known as the **numerical** response and the behaviourally-based **functional** response. The **numerical** response represents those changes in numbers which occur through time in a population in response to various environmental influences, both biotic and abiotic. The **functional** response is a change in value of a rate process in response to changes in some feature of the environment — it is not time defined. Figure 43 gives examples of the nature of the numerical and functional response for two universal ecological processes, predation and movement. In the case of predation, the numerical response is the way in which the numbers of predators respond to the numbers of prey, which relationship can be inferred from examination of time-dependent plots of the two. The functional response is the relationship connecting various

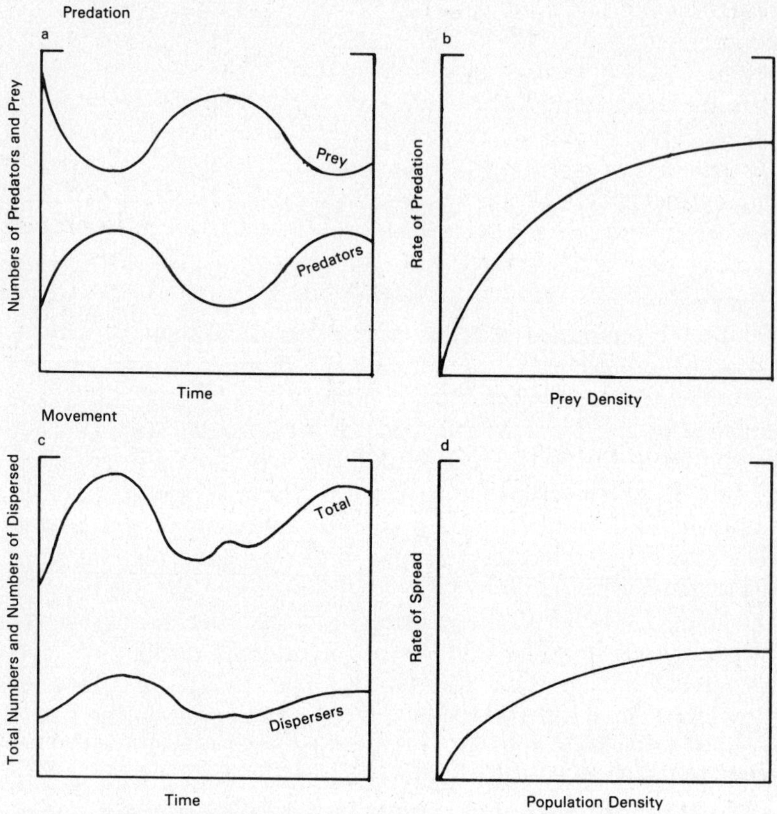

Fig. 43 Examples of plots of the numerical (a and c) and functional (b and d) responses within the processes of predation and movement

key processes within the predation act with some quality of the predator or prey; in the simple case shown the relationship is between the number of prey captured per unit time and the overall density of prey in the environment. For the movement process, the numerical response might be the relationship between the numbers of dispersers in a population of some animal and the overall numbers of animals in the population; once again the relationship can be inferred from examination of time-dependent graphs. The functional response in this instance, as in the case of predation, is a relationship between some characteristic rate of the ecological process and some other feature of the organism concerned. In the example shown, the functional response is that connecting the rate of spread against the numbers of animals present in the population as a whole. The curves shown in figure 43 are theoretical only and the processes involved will be discussed in more detail later in the chapter.

This chapter briefly examines a small selection of the many "simple" algebraic models of predation, competition and movement. These have a mathematical elegance and tractability which have made them popular tools for mathematical ecologists. They have led to considerable insights into how ecological processes operate and have a vital place in current ecological theory and practice. Many working ecologists have found, however, that these algebraic representations of processes are so simplified as to make their use limited. This led to the development of **component models** of key processes, where the process is subdivided into what the researcher sees as its component parts. Each part is modelled separately and the whole makes a more complex but, presumably, more realistic and flexible representation of the overall process. The latter part of this chapter contains detailed accounts of such a components analysis approach to predation and movement.

7.1 Albegraic Models of Ecological Processes

The starting points for most algebraic representations of ecological processes are the simple models given in table 17 or others of equal simplicity that show how a single species population grows under specified conditions.

7.1.1 Predation

The simplest and probably the earliest model of the **predation** process was that of Volterra (1926) which, in Roughgarden's version (1979) is:

$$\frac{dV}{dt} = rV - (aV)P \tag{1}$$

$$\frac{dP}{dt} = b(aV)P - mP \tag{2}$$

where V represents the numbers of the prey species, P the numbers of predators, r the rate of increase of the prey species, a the rate of predation, b the rate at which predators convert prey individuals into "new" predators, and m the death rate of predators. The model of course, is in the form of two simple differential equations (see 5.1) and is based on simple exponential growth or decline of the separate species populations. In the absence of predators (P = 0), equation 1 becomes:

$$\frac{dV}{dt} = rV$$

and, in the absence of prey (V = 0), equation 2 bccomcs:

$$\frac{dP}{dt} = -mP$$

Of course, most populations of prey species will be regulated in number, not only by the predator species, but also by a variety of other factors. The logistic rather than the exponential model can be readily used as the basis for the predation equation by simply substituting, as follows:

$$\frac{dV}{dt} = r(1 - \frac{V}{K})V-(aV)P \qquad (3)$$

where K is the "carrying capacity" of the prey population but all other symbols are as before. The equation for the predator population remains as in (2).

Many further additions can be made to incorporate time lags, lower limits below which predation is ineffective, predator satiation and density-independent mortality, to name but a few. May (1981c) and Roughgarden (1977) deal with many of these at length, together with mathematical treatments of the models' behaviour. Most analyses have focussed on the range of values for a, b, r, m, and K, for which stable populations of both interacting species, are possible. An example of the behaviour of the pair of equations (2) and (3), taken from Roughgarden (1977), is given as figure 44. The term (aV) in equations (1) and (3), of course, represents the behavioural interaction of the predator with the density of it prey and this is the functional response of predators to prey. In the simple Volterra models a is a constant but, as will be seen in 7.2.1, this can be substituted by a more or less complex function to increase the realism of the model represenation.

An alternative foundation for much later work was provided by Nicholson and Bailey (1935), who proposed a basic model for the interaction of parasitoids with their insect hosts — a rather special case of predation — and produced a model which Holling (1959), Hassell (1978) and many others have built upon.

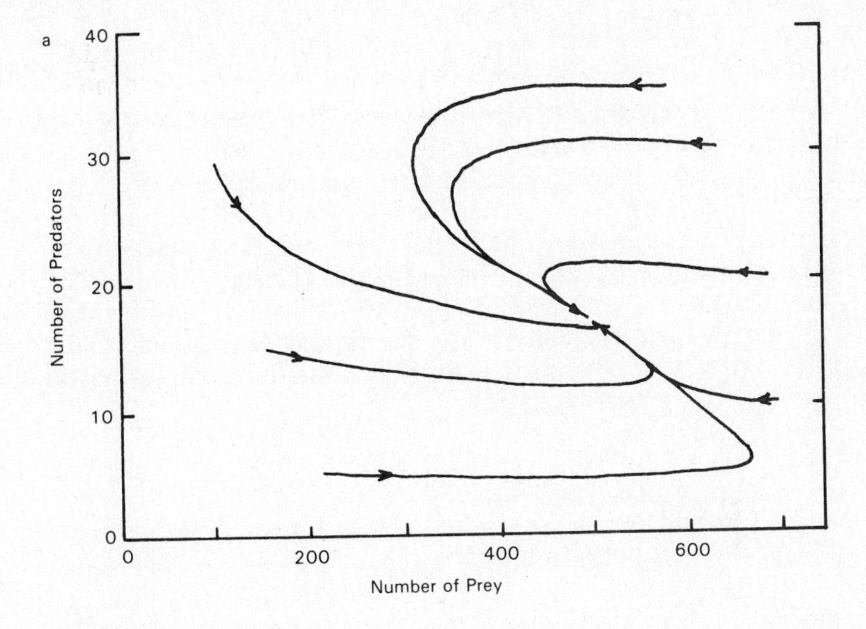

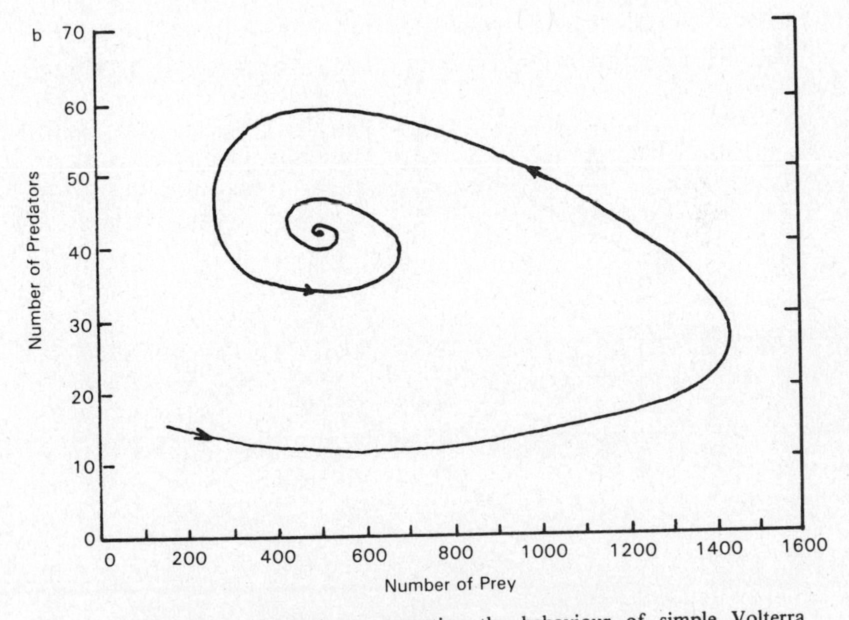

Fig. 44 Phase plane trajectories representing the behaviour of simple Volterra predator-prey equations. a. Parameters $r=0.5$, $a=0.01$, $b=0.02$, $d=0.1$ and $K=750$ and the trajectories approach the equilibrium point directly. b. Parameters $r=0.5$, $a=0.01$, $b=0.02$, $d=01$ and $K=3000$ and the trajectory spirals into the equilibrium point (after Roughgarden 1979)

$$V_{t+1} = \lambda V_t e^{-AP_t} \quad (4)$$
$$P_{t+1} = V_t(1-e^{-AP_t}) \quad (5)$$

where λ is the discrete rate of increase of the prey, A is the so called area of discovery of the predators for the prey, e represents the base of natural logarithms and other symbols are as before. Note that this time the equations of change are written in difference form. The area of discovery, A, is a measure of the searching efficiency of the predator. The form of the equation, based as it is on the Poisson distribution (5.4), assumes that predatory-prey encounters are distributed at random. Once again it will be obvious that the prey equation collapses to the exponential growth equation:

$$V_{t+1} = \lambda V_t$$

where no predators are present.

The Nicholson-Bailey equations, like those of Volterra, have been explored mathematically by many authors. As before investigations have centred on the stability properties of the model, which are summarized in figure 45 (after Hassell 1977).

The functional response in this model is contained in the (AP_t) term and it was from this actual formulation that C.S. Holling developed his first component model (see 7.2). Hassell

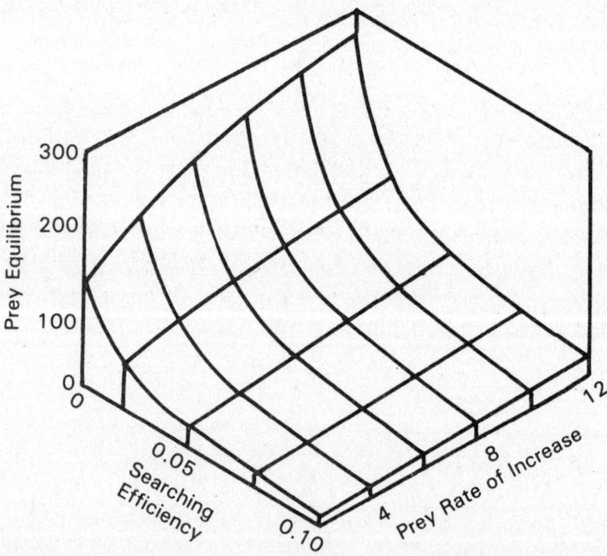

Fig. 45 Three-dimensional graph showing the interrelationships among the equilibrium level of prey, the searching efficiency of predators (A) and the rate of increase of the prey as predicted by the equations of Hassell (after Hassell 1977)

and his co-workers have incorporated terms to account for mutual interference between searching predators, refuges for prey and predator aggregation, and these are discussed at length in Hassell's 1978 book.

7.1.2 Competition

The process of **interspecific competition** has seen a very similar set of developments to those described for predation.

The simplest competition equations are the so called Lotka-Volterra equations, based upon the differential equation for logistic population growth. Again following Roughgarden (1979):

$$\frac{dN_1}{dt} = r_1 \left[\frac{K_1 - N_1 - \alpha_{12} N_2}{K_1} \right] N_1 \qquad (6)$$

$$\frac{dN_1}{dt} = r_2 \left[\frac{K_2 - N_2 - \alpha_{21} N_1}{K_2} \right] N_2 \qquad (7)$$

where N_1 and N_2 are the numbers of each species present, r_1 and r_2 their respective rates of increase, K_1 and K_2 the two carrying capacities involved; and α_{12} and α_{21} are the so called **competition coefficients** of one species on the other, expressing the magnitude of the competitive effect in each direction. Once again, analyses using this model have been centred on its stability properties and, traditionally, the approach has used graphical techniques which are summarized, following Roughgarden (1979), in the series of graphs given in figure 46.

The most important simple extension to this model is due to Pianka (1981), who notes that any particular species will compete, not with one but with several other species. This is modelled by the equation:

$$\frac{dN_i}{dt} = r_i \left[\frac{K_i - N_i - \sum_{j=1}^{n, \; i \neq j} \alpha_{ij} N_j}{K_i} \right] N_i \qquad (8)$$

This says that for the ith species out of n in a competing complex, the reproductive rate, r_i, of the species concerned will be reduced in proportion to the abundances of all the other com-

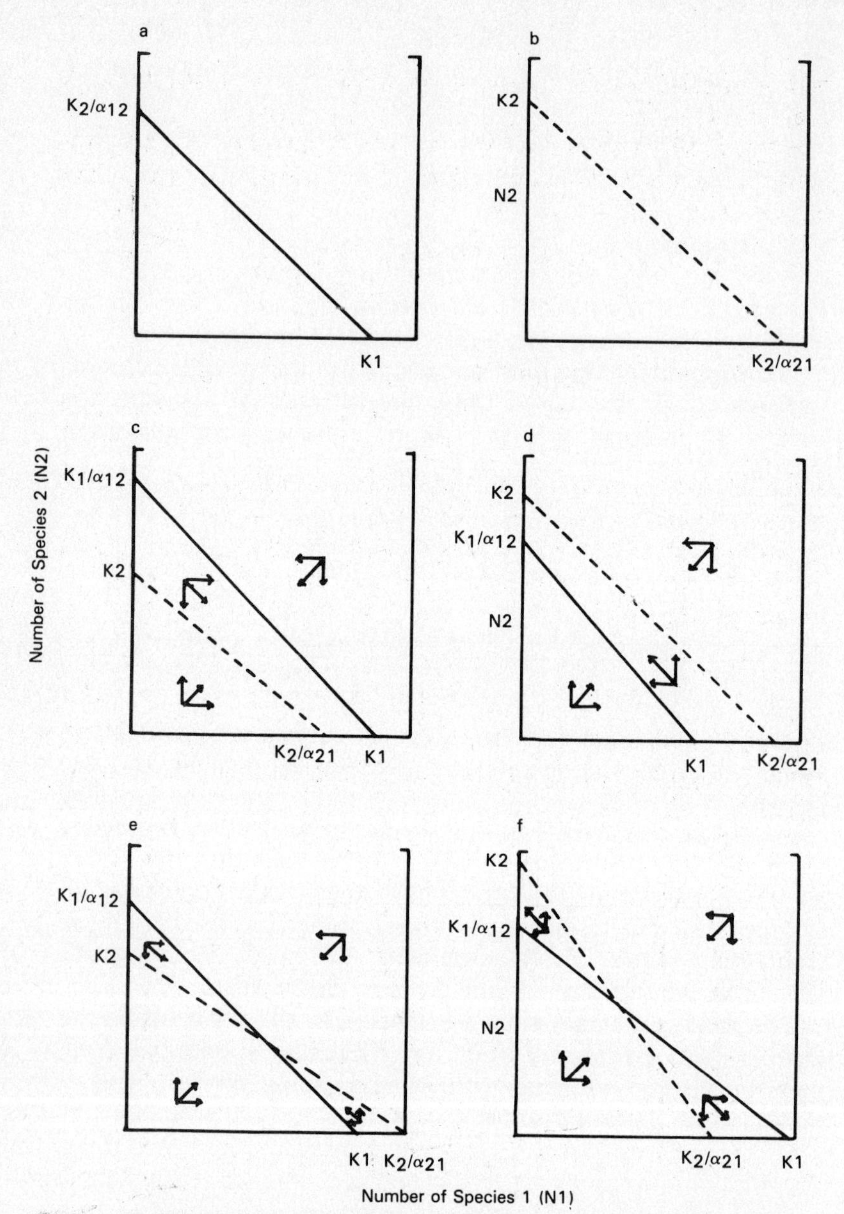

Fig. 46 Graphical analysis of competition equations as presented by Roughgarden (1979). a. The line along which $dN_1/dt = 0$. b. The line along which $dN_2/dt = 0$. c. The case where $K_1 > K_2/\alpha_{21}$ and $K_2 < K_1/\alpha_{12}$ – there is no stable equilibrium for the pair of species and species 1 always wins. d. The case where $K_2 > K_1/\alpha_{12}$ and $K_1 < K_2/\alpha_{21}$ – again there is no stable equilibrium and species 2 always wins. e. The case where $K_2 < K_1/\alpha_{12}$ and $K_1 < K_2/\alpha_{21}$ – there is a two-species equilibrium point (that is the two trajectories do intersect), but this is an unstable equilibrium, minor deviations from it leading to the extinction of one or other species. f. The case where $K_2 > K_1/\alpha_{12}$ and $K_1 > K_2/\alpha_{21}$ – there is a globally stable, two species equilibrium point, deviations from which are followed by a return to the equilibrium point

peting species, each operating through its own competition coefficient, α_{ij}.

Hassell (1976) gives a difference equation version of the competition equations which parallels, in many ways, his treatment of predation. His equations are:

$$N_1(t+1) = \lambda_1 N_1(t) \left\{1 + x_1 [N_1(t) + \alpha_{12} N_2(t)]\right\}^{-y_1} \quad (9)$$

$$N_2(t+1) = \lambda_2 N_2(t) \left\{1 + x_2 [N_1(t) + \alpha_{21} N_1(t)]\right\}^{-y_2} \quad (10)$$

where $N_1(t+1)$, $N_1(t)$, $N_2(t+1)$ and $N_2(t)$ represent population levels of species 1 and 2 at times t and t+1 respectively, λ_1 and λ_2 are the rates of increase of the species, and x_1, x_2, y_1 and y_2 are fitted constants. Both these equations are based on the single species form:

$$N(t+1) = \lambda N(t)[1 + xN(t)]^{-y} \quad (11)$$

of Hassell (1975), which can be made to represent both scramble (each competitor getting a little of some limiting resource) and

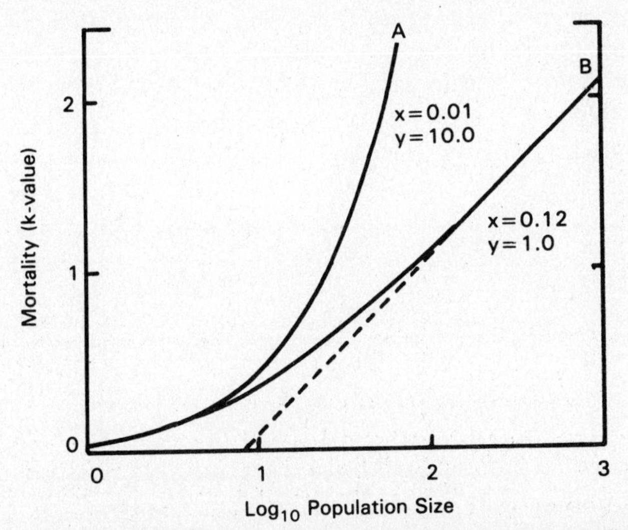

Fig. 47 The relationships between mortality levels and population size predicted from Hassell's (1975) intraspecific competition equations under two separate sets of parameter values: A. mimicking "scramble" and B. "contest" competition (after Hassell 1975). The use of the "K-value" as a measure of mortality follows the usage of Varley and Gradwell (1970) and is explained further by Hassell (1975)

contest (each competitor getting all or none of a particular resource) competition within a species. Figure 47 summarizes these effects and demonstrates the role of the constants x and y in the dynamics of the equation.

7.1.3 Movement

Algebraic models of **movement** are usually more complex than those of predation and competition and they are much less prominent in the literature. Most have been based on the classical partial differential equations of the diffusion process which are beyond the scope of this work. Pielou (1977) provides a good introduction to the subject.

Gadgil (1971) wrote a pair of difference equations based on the logistic formulation (see table 8):

$$N(t+1) = N(t) + rN(t)\left[\frac{K - N(t)}{K}\right]$$

He then postulated that for a pair of populations, N_1 and N_2, of the same species, a proportion p_1 would move from population 1 to 2 and a proportion p_2 from population 2 to 1 in any time interval. Thus in any time interval each population would have a proportion of "dispersers", p, and a complementary proportion, (1-p), of "stay-at-homes". Assuming independent logistic growth within each population, he was able to write:

$$N_1(t+1) = (1-p_1)N_1(t) + rN_1(t)\left[\frac{K_1 N_1(t)}{K_1}\right] + p_2N_2(t) + rN_2(t)\left[\frac{K_2\ N_2(t)}{K_2}\right] \quad (12)$$

and

$$N_2(t+1) = (1-p_2)N_2(t) + rN_2(t)\left[\frac{K_2\ N_2(t)}{K_2}\right] + p_1N_1(t) + rN_1(t)\left[\frac{K_1 - K_1(t)}{K_1}\right] \quad (13)$$

Building on this basic model, Gadgil examined the effects of introducing a special mortality factor during dispersal, variation in the size of K through time, and the generalization to n populations with competition among them. He used his results to make a series of statements about the evolutionary advantages of various strategies of dispersal.

7.2 Component Models of Predation and Animal Movement

In his approach to the study of the process of predation, C.S. Holling developed what has become known as the **experimental components** approach. In essence this was an investigation of the functional response part of the process defined above, by dividing the response into a number of basic and subsidiary components. The basic components were seen as being nearly universal, operating in virtually all examples of the process, whereas other components, added either at the same primary level or at subsidiary levels, characterize more specialized forms of the process. This operation of identifying the structure of the process is of general applicability; we shall see how Holling applied his ideas to predation and how I attempted to apply some of the same principles to a modelling study of the movement process.

7.2.1 Predation

In a series of papers beginning in 1959, C.S. Holling examined the predation process in detail, using the philosophy and techniques of the experimental components approach outlined above. He defined ten **components of predation** at the primary level of resolution and built a series of simulation models, starting with a model containing relatively few components and adding more in subsequent models. Table 30 shows the ten components identified by Holling and indicates whether these are **basic**, that is, whether they would be expected to occur in all instances of the predation process that can be imagined; or **subsidiary**, that is, absent in some cases. The first line of the body of table 30 indicates the various paths that can be taken through this sequence of components, the various combinations of which can represent any instance of predation. Holling's first model of

Table 30. The components of predation (modified from Holling 1966)

	POSSIBLE COMPONENTS OF FUNCTIONAL RESPONSES TO PREY AND PREDATOR DENSITY									
	Rate of Successful Search	Time Exposed	Handling Time	Hunger	Learning by Predator	Inhibition by Prey	Exploitation	Interference between Predators	Social Facilitation	Avoidance Learning by Prey
BASIC OR SUB-SIDIARY	Yes →	Yes →	Yes →	Yes / No	Yes / No	Yes / No	Yes →	Yes →	Yes / No	Yes / No
MODELS TO DATE	Holling 1959									
	Holling 1966									
	Holling 1965									
							Griffiths and Holling 1969*			
										Dill 1973*

*The published papers describe submodels conceived as part of the full model initiated by Holling.

predation, the so called **disc equation model**, contained only the first three universal components of the process, namely the rate of successful search by a predator, the time of exposure by the prey, and the handling time. This last component is a measure of the amount of a predator's time used up in the actual act of predation, and which is therefore unavailable as time for searching for further prey. The model built using these three components was the subject of Hollings' 1959 paper; in subsequent papers (1965, 1966) he added further components on top of the basic ones. This account will examine the first two of these models to convey the flavour and technical aspect of the approach pioneered by Holling: the **disc equation model** and the **invertebrate response model** (1966) cover four of the ten basic components of the predation process. Holling's **vertebrate response model** (1965) adds a further two and all but one of the other four components have been treated by subsequent workers. The exploitation and interference components were the subject of a submodel of the competitive part of the predation process built and presented by Griffiths and Holling in 1969. The component, avoidance learning by the prey, was also elucidated and presented as a submodel by one of Holling's students, L.M. Dill, using some elegant techniques of mathematics and observations using fish prey and predators as subject matter.

The approach developed by Holling is applied throughout this sequence of papers and consists of an effective combination of laboratory experimentation, field observation and mathematical or numerical modelling. Holling used a philosophy closely similar to that outlined in chapter 3, where the basic cyclical nature of the modelling process was identified. In the case of the experimental components approach under consideration here, this consists of step by step additions to the model, each step being tested by carefully designed experiments before progression to the next step, Holling was one of the first ecologists to realize that a general and useful approach to the study of an ecological process could be made, not by choosing one instance of the process from the large array of such processes available in the natural world and studying that in detail, but instead, building an abstract model of the process and testing it out against any instance of the operation of the process that was convenient for experimental purposes. He and his co-workers and students developed the idea that what was important was, not so much the actual values to be placed on constants in the mathematical formulation, but the nature of the formulations themselves, the functional forms employed and the overall behaviour or set of behaviours of the systems model.

It must be observed, however, that most of the experimentation involved in the sequence of models developed by Holling was carried out in the laboratory; only in very few instances were field data used. I return to this point below.

The Disc Equation Model

In the first of Holling's modelling attempts, he argued that the functional response of predators to the density of their prey would be governed by three components, which he suggested are the most universal of the characteristics of the predation process:

the rate of successful search for predators,
the time of exposure of the prey to the predators, and
the time required for the actual handling of a prey by the predator.

To combine these features of the process, Holling built a mathematical model based upon earlier ideas of Nicholson and Bailey (1935), and the logical development he used is set out in table 31. Wherever possible I shall retain the symbols used by Holling. The functional relationship obtained describes a smoothly rising curve which levels off as the prey density increases. Holling labelled this relationship the **Type II** func-

Table 31. The development of Holling's disc equation model (Holling 1959)

Let N_A be the number of prey captured
N_O by the density of prey individuals
T_S be the time available to the predator for searching.
Then, most simply, N_A will be proportional to $T_S N_O$
i.e. $N_A = a.T_S.N_O$ (1)
where a is a constant, the "rate of discovery" of prey items by the predator.
Now, if T_H is the time taken to "handle" one prey item, the total time (T_T) budget of the predator can be stated as:

$$T_S = T_T - T_H.N_A$$

This expression can be substituted into equation (1) to obtain:

$$N_A = a(T_T - T_H.N_A)N_O$$

Multiplying out:

$$N_A = T_T.a.N_O - a.T_H.N_O.N_A$$

whence:

$$N_A + a.T_H.N_O.N_A = T_T.a.N_O$$

and:

$$\boxed{N_A = \frac{T_T.a.N_O}{1 + a.T_H.N_O}}$$

which is the so called disc equation.

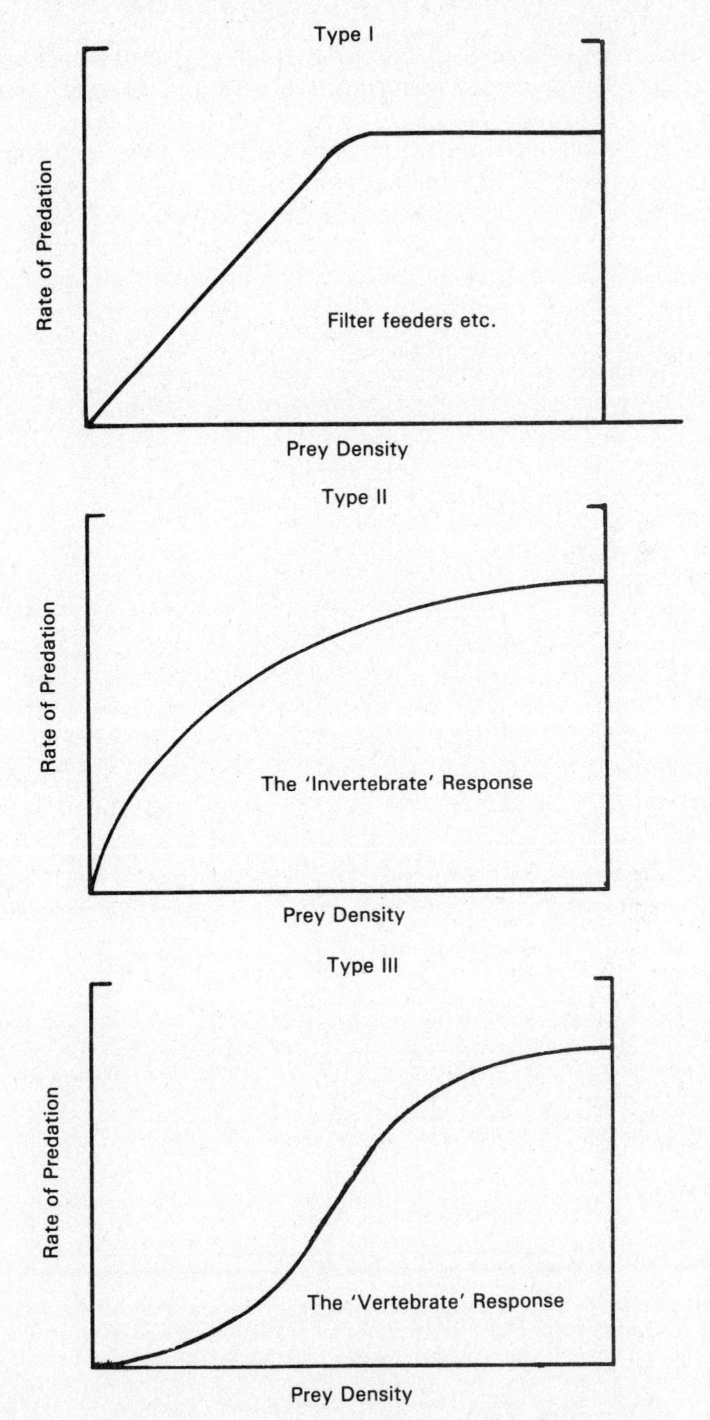

Fig. 48 The three basic types of functional response (after Holling 1959)

tional response, and identified two other possible classes of response for the relationship, the **Type I** and **Type III** respectively. Figure 48 illustrates these three classes of response. Having obtained this expression, Holling's next step was to check it against experimental results. In looking around for a suitable example of predation for this purpose, he lit upon the idea of a physical simulation exercise by which he could imitate the process, but keep it sufficiently simple so that the results could be compared directly against those predicted from the, admittedly, simplistic disc equation model. He spread a number of sandpaper discs on a table, each disc acting as a "prey" individual; his "predator" was a blindfolded assistant who searched for the "prey" by tapping with a finger on the table top. Holling was able to calculate all the parameter values required by the model and then to follow the behaviour of the model, the

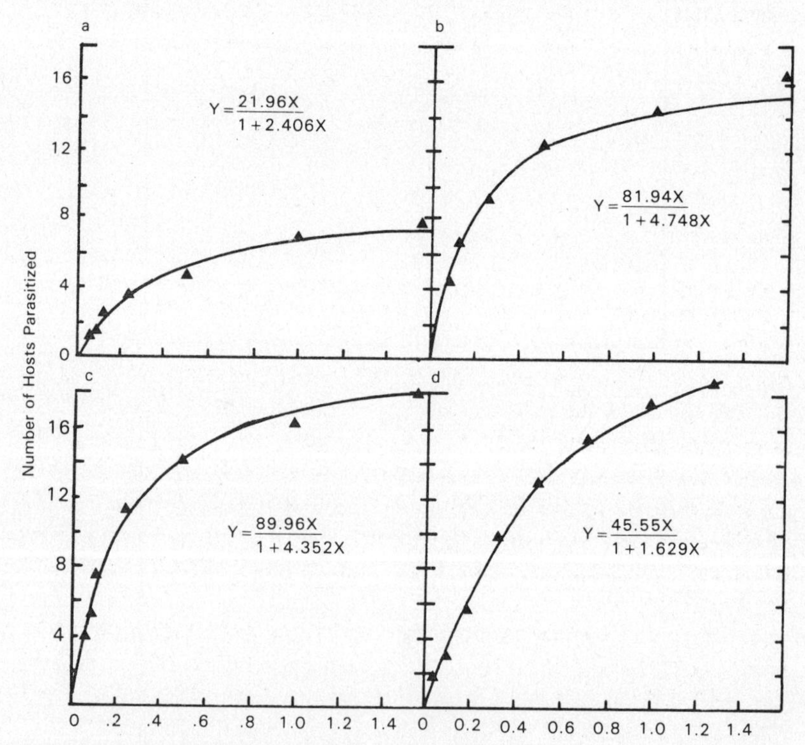

Fig. 49 Type II functional responses, describing the searching of the parasitoid *Dahlbominus fuliginosus* for cocoons of the sawfly *Neodiprion sertifer*. The four sets of points represent results from trials carried out under different conditions: a, b and c were maintained at 16, 20 and 24°C respectively, with different cocoon densities achieved by using different cage sizes, and d at 24°C, with cocoon densities manipulated in cages of the same size. The curves represent the predictions of the disc equations shown (after Holling 1959)

predictions of which could be compared with that observed in his experiment. The predictions of the model, developed using these three components only, were closely similar to those obtained in the experimental situation; in addition, the simple model presented was able to describe adequately the situation observed in several instances, where the response of insect parasitoids (a form of predator where the prey is used as an oviposition site and the resulting predator larvae eats the prey species from the inside) was measured. Figure 49 shows some of the results used by Holling and described in his 1959 paper, and compares these with the predictions of the disc equation model.

Although elegant and very satisfactory as a description of some simple forms of predation, notably those observed in the interaction of insect predators and their prey, this model fails to take account of several other components which, it is to be supposed, are involved in the predation process at all levels. Holling considered the most important of these to be the degree of **hunger** of the predator, which changes through time and with respect to the feeding history of the organism. His next essay in model building, therefore, attempted to incorporate this additional component into the model containing the three basic components just described.

The invertebrate response model

In a paper published in 1966, it was Holling's intention to add the component, hunger, to the three used in his disc equation model, thus obtaining a more general expression or set of expressions for representing the predation process. In order to do this, however, he had to redefine the three general components incorporated in the disc equation and break them down into the constituent subcomponents listed in table 32. The 1966 paper then examined the way in which levels of hunger affect each of the ten subcomponents involved in this more detailed examination of the three basic components of his earlier model.

Throughout this piece of work Holling used a praying mantis, *Hierodula crassa,* preying upon either fruit flies or house flies as his experimental animal. This particular species was most suitable for Holling's purposes, though he was quick to point out that the predator and prey species were chosen for convenience, and that the model developed was intended to have a general relevance beyond its application to the particular instance studied. The model produced was claimed by Holling to

Table 32. The subcomponent structure of the three qualities identified by Holling as basic to the predation process

1. Rate of successful search (a)
   (a) the reactive distance of the predator for prey (r_m)
   (b) the speed of movement of the predator (VD)
   (c) the speed of movement of the prey (VY)
   (d) the capture success (SC) (i) recognition success (SR)
   (ii) pursuit success (SP)
   (iii) strike success (SS)

2. The time prey are exposed to predators (T_S)
   (a) the time spent in activities not related to feeding (TN)
   (b) the time spent in feeding activities (TA)

3. The time spent handling each prey (T_H)
   (a) time spent pursuing and subduing each prey (TP)
   (b) time spent eating each prey (TE)
   (c) time spent in "digestive pause" after prey is eaten (TD)

fit all the cases of invertebrate predation examined to that date, and for this reason was entitled the **invertebrate response** model. The validity of this claim of generality will be examined later.

Before Holling could examine the effect of hunger on each of the components listed in table 32, it was necessary for him to work out an operational measure of just what hunger was in his experimental animals, and to determine how it could be measured at any particular point in time. He was able to show that a general relationship could be arrived at for mantids, relating the weight of prey that a particular individual could be made to eat at any time and the period over which it had been deprived of food previously — in other words the amount of food required to satiate the animal under a variety of different conditions of hunger. Table 33 illustrates this relationship, showing the form of the general equation which Holling was able to fit to these results. Basically the equation describes a negatively accelerating, rising curve, and involves the maximum capacity of the gut of the predator and the rate of disappearance of the food from the predator's gut. Holling manipulated this equation algebraically in the fashion set out in table 33 to obtain an alternative expression for hunger, where the time element involved was expressed not as "time deprived of food" but in terms of unspecified time. This form was more easily integrated into some of the expressions for other components that he developed later in his 1966 paper (see below). Holling examined a number of sets of data for different animals and concluded that this equation, relating the weight of food eaten to the time of deprivation, was of a general form which could be used to

Table 33. Derivation of the hunger equation for the "invertebrate response" model

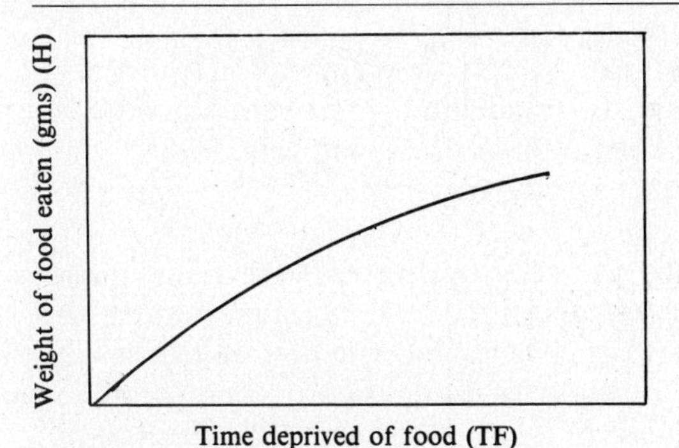

This relationship is of a general form

$$H = HK\,(1 - e^{-AD(TF)}) \qquad (1)$$

where H and TF are as in the figure, HK is the maximum capacity of the gut and AD is the rate of disappearance of food from the gut.
We can rearrange this to obtain an expression for TF

i.e. $$TF = \frac{1}{AD} \ln\left(\frac{HK}{HK-H}\right) \qquad (2)$$

We may wish to express this not in terms of time deprived of food (TF) but in unspecified time (T) such that H = HO at T = 0.
Thus, if TF = TFO when H = HO, we obtain an expression for the hunger (H), T units of time later, i.e. at TFI where

$$TFI = TFO + T \qquad (3)$$

From equation (2) we can write

$$TFI = \frac{1}{AD} \ln\left(\frac{HK}{HK - H}\right) \text{ and } TFO = \frac{1}{AD} \ln\left(\frac{HK}{HK-HO}\right)$$

which can be substituted into (3) and rearranged to give:

$$T = \frac{1}{AD} \left[\ln\left(\frac{HK - HO}{HK - H}\right) \right]$$

which again rearranges to:

$$H = HK + (HO - HK)\,e^{-AD(T)} \qquad (4)$$

– the hunger equation in terms of unspecified time.

quantify levels of hunger for all predators that he examined. Having described these experiments, Holling thereafter used the weight of food required to satiate animals as synonymous with level of hunger of that animal. Having defined and established his criteria for the hunger process, Holling proceeded to examine the effects of this component on the other three major components of predation that he had dealt with previously. His approach to this modelling procedure was somewhat unconventional and in many ways characteristic of the experimental components philosophy. He took each subcomponent and looked at the biology of the process involved, going through a modelling and experimentation procedure for each; he then attempted a synthesis, bringing all of these equations together in an overall model of the predation process.

Holling first examined the relationship between hunger and the **rate of successful search.** This involved separate examination of the subcomponents of the search rate, the first of which was the size of the **area of reaction** of the predator for prey. This area of reaction is the space in front of the predator, the mantid in this case, within which a reaction of some sort to a prey item could be expected. Holling pointed out that in other sorts of predators this reactive field might be more than two-dimensional, but his analyses were restricted to this simpler case. At the start of his investigation of this relationship, Holling presented mantids of known hunger levels with flies at different distances directly in front of them and observed the maximum distance away from the mantid that produced a response. He divided the response into two forms, that which produced a physical move on the part of the mantid, the **move** response, and that which simply caused the mantid to move its head, the **awareness** reaction. His subsequent analyses were primarily concerned with the move response, although he used the same sorts of expressions and models for both levels of response. Table 34 lays out his description of the relationship between the maximum distance of reaction and the level of hunger, using the expression that he had worked out previously, connecting hunger with the time of deprivation of food (see table 33). He was able to show that there was a threshold level below which no response by the predator could be obtained at all and above which there was an increasing, maximum distance of reaction with increase in the period of deprivation of food. When this period of deprivation of food was converted into hunger using the equation referred to earlier, the relationship with the reaction distance became linear. This permitted Holling to describe the relationship in relatively simple terms mathematically and his procedures are shown in table 34.

Having linked the maximum distance of reaction of the predator to the level of hunger, Holling complicated his analysis by pointing out that the linear reactive distance in front of the mantid was not so important as the **area** of reaction around the

Table 34. Calculation of the effects of hunger on the reactive distance of a predator (after Holling 1966)

Let r_m be the maximum distance of reaction and H be the level of hunger of the predator, defined in terms of the period deprived of food (TF) by the equation

$$H = HK\,(1 - e^{-AD(TF)})$$

where HK and TF are as defined in fig. 19
then plotting:

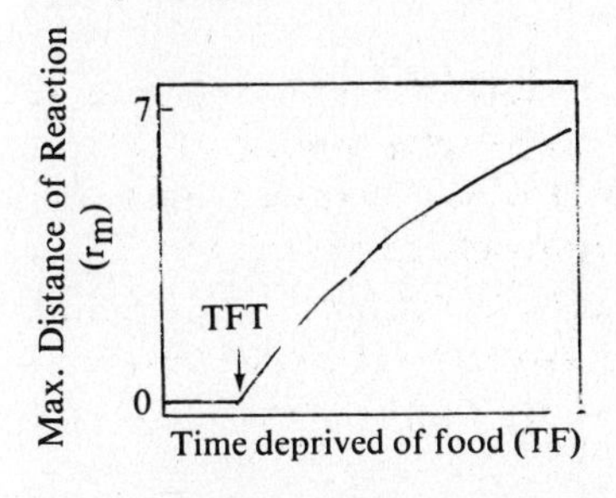

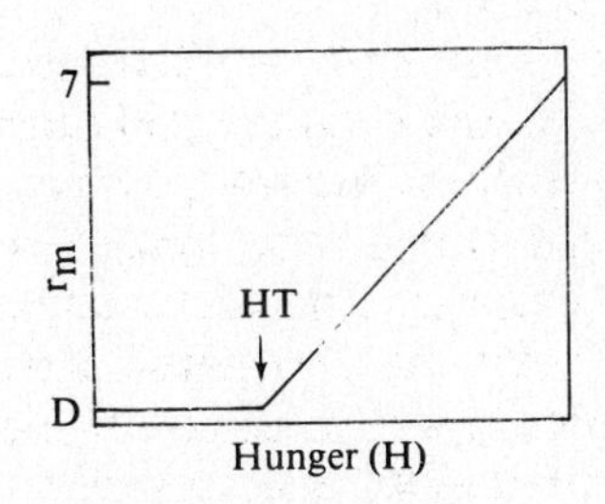

HT is the threshold value of hunger below which no reaction occurs.
Transforming TF to H using the earlier equation this relationship is linearized, whence:

$$r_m = GM\,(H - HT) \qquad H > HT$$
$$r_m = 0 \qquad H \leq HT$$

where GM is the slope of the line.

We can expand this by substituting for H:

$$r_m = GM\,(HK - HT - (HK)e^{-AD(TF)}) \qquad TF > TFT$$
$$r_m = 0 \qquad TF \leq TFT$$

where TFT is the TF value when H = HT

full 360 degrees field surrounding the predator's anterior end. Mantids in fact are able to see most of this area. Holling plotted this field of awareness for mantids at different stages of satiation, with a technique of experimentation similar to that used in measuring reactive distances in front of the predator. His results are shown in the first part of figure 50. He then needed to obtain an expression relating the shape and area of this reactive field to the level of hunger of the predator. His derivation of this relationship is set out in table 35 and takes as its starting point the definition of the minimum visual angle, α, as the basic parameter to which the predator is responding, rather than the actual distance of the prey from the predator. In this connection Holling reasoned that distance would not be universally appropriate for measuring reactive behaviour, because prey items would differ in size and a large item at a certain distance from the predator would be likely to produce more response than a small item at the same distance. He

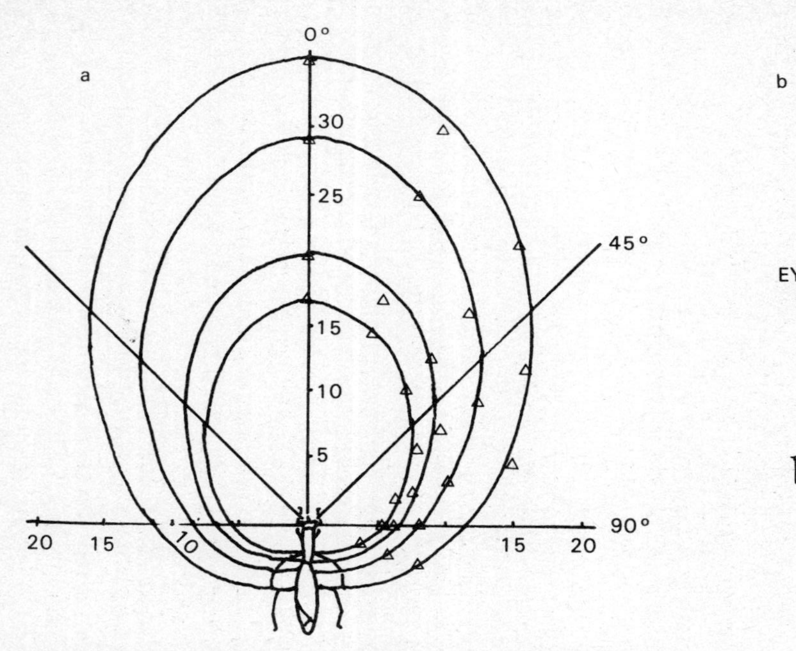

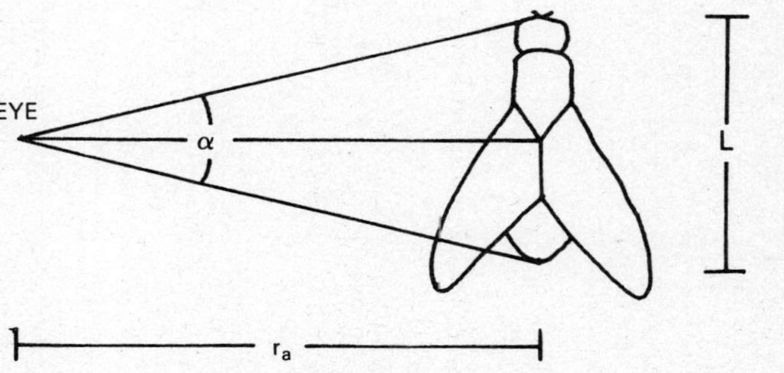

If r_a is the distance from the mantid to the edge of its field of reaction at any point, then the area (A) of this field can be obtained by polar integration with respect to the angle subtended with the axis of the body (θ):

$$A = \int_{\theta=0}^{\theta=\pi} r_a^2 . d\theta \qquad (1)$$

where θ is expressed in radians.

$$\alpha = 2 \arctan L/2r_a \qquad (2)$$

whence

$$r_a = L/[2 \tan(\alpha/2)] \qquad (3)$$

when α is small and $\tan\alpha$ tends to α

$$r_a = L/\alpha \qquad (4)$$

Fig. 50 Development of terms within the invertebrate response model of Holling: A. the reactive field of searching mantids; B. the angle subtended by a prey item on the eye of the predator (after Holling 1966). In a the open triangles represent data points.

Table 35. Derivation of the relationship between the area of the reactive field of a predator and its level of hunger based on the data and relationships given in table 33

1. Linearize the expression for the minimum visual angle (α)

   i.e. $\alpha = \alpha_o + m\,\theta^2$

   where θ is the angle to the axis of the body of the predator, α_o is α when $\theta = 0$ and m is the slope of the linear relationship.

2. We expand m because the relationship between α and θ is a function of the structure of the eye such that α_i/α_o is a constant (where α_i is α when $\theta = i$).

   Whence: $m = \alpha_o C$

   Where C is an independent constant that describes the way the orientation of the visual elements in the eye change from anterior to posterior.

3. Substituting: $\alpha = \alpha_o + \alpha_o\, C\, \theta^2$

4. Where α is small we can substitute this in equation 4 of fig. 50.

   i.e. $r_a = L/(\alpha_o + \alpha_o C\theta^2)$
   $= L/\alpha_o(+ C\theta^2)$

   But $L/\alpha_o = r_o$ (from (4) in fig. 19)

   then $r_a = r_o/(1 + C\theta^2)$

   which is a measure of the way r_a and, hence, the shape of the reactive field varies.

5. To move from this to a measure of the **area** of the field we substitute in equation (1) of fig. 46

   $$\text{i.e.} \quad A = \int_{\theta=0}^{\theta=\pi} [r_o/(1 + C\theta^2)]^2 d\theta$$

   where θ is converted to radians

   $$\text{whence} \quad A = r_o^2 \int_{\theta=0}^{\theta=\pi} d\theta/(1 + C\theta^2)^2$$

   Integrating using standard methods

   $$A = \frac{r_o^2}{4\sqrt{C}} \left(\frac{\pi\sqrt{C}}{1 + C\pi^2} + \text{Arctan}\, \pi\sqrt{C} \right)$$

   π and C are constants so that this can be simplified to:

   $$\boxed{A = KA\,(r_o^2)}$$

   where KA is an areal constant characteristic of the particular predator.

6. Knowing the relationship between r_o and hunger (see table 33 where $r_m = r_o$) we can substitute to obtain the desired relationship

   $$A = KA\,[GM\,(H - HT)]^2 \quad H > HT$$
   $$A = O \quad H \le HT$$

   where all symbols are as used previously.

expressed this complication in terms of the minimum visual angle, then redefined this angle in terms of the position of the prey with respect to the main axis of the predator, developing the relationship shown in step 1 of table 35. The constant relating the minimum visual angle to the position of the prey vis-à-vis the predator was transformed to take account of the way in which the shape and receptive nature of the predator's eye changes from anterior to posterior. These transformations effectively linearized the relationship between the minimum visual angle and the angle of the prey with respect to the axis of the predator. Holling then substitued the expression for this angle into the equation developed in figure 50b, showing the relationship between the **reactive distance,** the **size of the prey** and the **minimum visual angle.** He used a simplification which assumed that the minimum visual angle would always be very small at the limit of awareness of predator for prey, and had a simple measure of the way the reactive distance and hence the shape of the reactive field varied from the front to the rear of the predator. He was able to move from this expression of the linear distance of awareness to a measure of the area of the reactive field, using a polar integration process employing standard methods and set out in step 5 of table 35. The expression this obtained for the area can be simplified to the equation shown in step 5 of the table, where there is a linear relationship between the area and the square of the reactive distance immediately in front of the prey, the position at which it is most easily measured experimentally. The constant in this relationship is an areal measure characteristic of the reactive field of the particular species of predator. From the work relating reactive distance and hunger set out in table 34, Holling was able to substitute for the reactive distance in this simple equation and obtain a relationship between the area of the reactive field and the level of hunger in the particular predator.

The next stage of the process was to examine **speed of movement** of the predator and of the prey with respect to the level of hunger in the predator. The speed of movement of the predator was assumed by Holling to be zero in the first instance, as this most matched the experimental situation which pertained in the mantid/fly system. The mantid is a "sit and wait" predator until it begins the actual pursuit of a particular prey item. The speed of movement of the prey however, must be taken into account, and Holling pointed out that prey activity would be affected mainly by the density of prey individuals in a particular area. He made a series of sets of observations on the number of

Table 36. The relationship between average prey velocity and prey density (after Holling 1966)

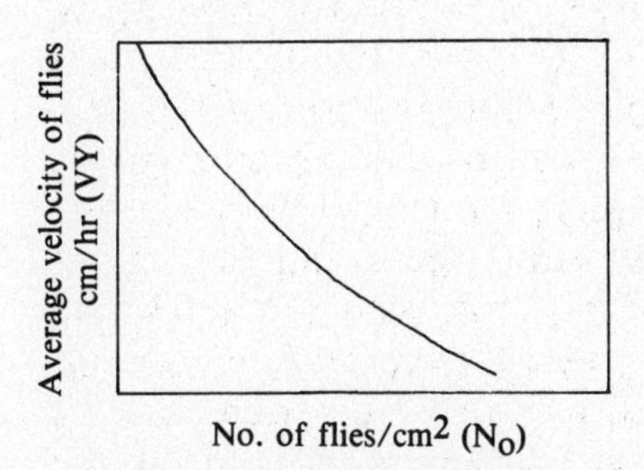

A suitable fitted equation is of the form

$$VY = (V_{max} - V_{min})\, e^{-d_V N_o} + V_{min}$$

where V_{max} is VY when $N_o = 0$
V_{min} is VY when $N_o = \infty$
for Holling's data this has parameter values such that:

$$VY = (315 - 92)\, e^{-139.3 N_o} + 92$$

contacts made between moving flies and a region immediately in front of the predator at a variety of different prey densities. The results obtained are shown in table 36 and were translated into an empirical equation in the manner shown in table 37. Holling points out that this expression for the average velocity of his prey items is very specific to the experimental situation he chose to study and has little generality. He observes "the whole process of animal activity should therefore be investigated using the same kind of component analysis used here for predation". It was this suggestion in part that led to the attempt at modelling animal movement described in the second half of this chapter.

Returning to Holling's invertebrate response model, he completed his examination of the rate of successful search by looking at the subcomponent, **capture success.** He observed that a contact of predator with prey did not always lead to a successful predation event, and that a series of stages in the predator/prey contact would each have a different rate of success associated with it. In order to analyze this, Holling subdivided capture success further into **recognition success, pursuit success** and **strike success.** He examined each of these in turn and showed that **recognition success** could be measured as the ratio between the number of recognized encounters over total encounters, which in the case of the mantid was one. **Pursuit success** he defined again as a ratio, this time as the number of pursuits over total encounters, and he developed an expression for this pursuit

Table 37. Derivation of the equation for the speed of movement of the prey for a stationary predator (after Holling 1966)

Let VY be the speed of movement of the prey
r be the radius of the area of detection around the prey.

Thus for a walking prey the area "swept" in time t will be as:

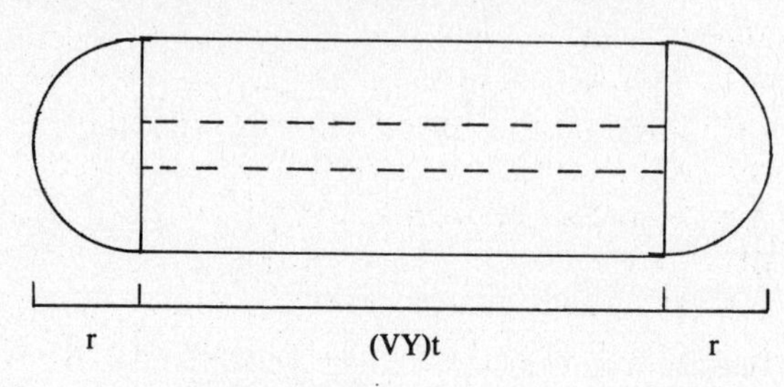

i.e. $2\,r\,(VY)\,t + \pi\,r2$

In a total area A_c with a prey density of N_o per unit all prey in the area will sweep jointly:

$$[2r\,(VY)t + \pi\,r^2]\,N_o\,A_c$$

Given that there is only 1 "contact" area A_c the density of contacts is $1/A_c$ then the number of times a prey enters the contact area, N_c, is:

$$N_c = [2\,r\,(VY)\,t + \pi\,r^2]\,N_o\,A_c\,(1/A_c)$$
$$= [2\,r\,(VY)\,t + \pi\,r^2]\,N_o$$

from whence $VY = (N_c - \pi\,r^2 N_o)/2rtN_o$

for Hollings mantids $t = 1$ hour and $r = 1.5$ cm,
hence $VY = (N_c - 7.069N_o)/3N_o$

Table 38. Derivation of the relationship between capture success and hunger for a predator (after Holling 1966)

Capture success (SC) consists of 3 stages:

1. Recognition success (SR), i.e. no. of recognized encounters/total encounters.
2. Pursuit success (SP), i.e. no. of pursuits/total encounters.
3. Strike success (SS), i.e. no. of strikes/total encounters.

For Holling's mantid/fly system:

$$SR = 1$$
$$SS = 0.630$$

SP in mantids is in fact the probability that a fly being stalked remains stationary during the time of pursuit (TP). If this is so then the probability of a fly not moving is e^{-mt} where m is constant and t is time. This is the "zero" frequency class of the Poisson distribution, i.e. assumes durations of movements are randomly distributed.
Whence:

$$I = ne^{-mt}$$

Where I is the number of periods of immobility lasting t time units and n is the total numbers of periods of immobility.

Therefore $SP = I/n$
or $SP = ne^{-mt}/n$
Substituting TP for t and rearranging:
$SP = e^{-m(TP)}$
Thus $SC = (SR)(SS)\,e^{-m(TP)}$

success in terms of the probability of a particular prey item remaining stationary through the duration of the pursuit event. The logic and algebra involved are laid out in table 38. He completed his examination of the different segments of the capture success component by looking at **strike success**, and was able to show that for the mantid at least strike success was constant and independent of the level of hunger, having an average value of 0.63 measured over some six hundred odd replicates. The three success terms were combined to obtain the overall capture success, as shown in the later part of table 38.

The next objective in Holling's development of this particular model was to examine the second major component of the predation process, the **time prey were exposed to predators.** He recognized two subcomponents of this quantity, namely the time spent in **activities not related to feeding** and the time spent in **feeding activities.** As in the case of his treatment of the speed of movement of the prey, he adopted an empirical approach, in this instance pointing out that, for the experimental situation that he was examining, both of these quantities were constant, the period of feeding of predators being restricted to sixteen hours of light with an eight hour period of non-feeding during the hours of darkness. His experimental animals were not permitted to spend time in activities other than feeding during the period of daylight. It is this sort of sweeping assumption that detracts from the generality and utility of Holling's models as, particularly in field situations, they will be violated more often than not. The field applicability of Holling's work will be considered at the end of our treatment of his model.

The last major component Holling examined in this piece of research was described as the **time spent handling each prey.** He divided this into three subcomponents: the time associated with **pursuit** and the **subduing** of each prey item; the time spent **eating** each prey; and the time spent in **digestive pause** after a prey is eaten, during which the hunger level of the predator is such that no response to further prey items can be elicited in the predator. He developed a general expression relating **pursuit time** and the level of hunger, which is laid out in full in table 39. Once again Holling used an elegant combination of geometry and algebra to obtain the general relationship he sought. He was able to use a much simpler approach for the component relating to the time spent **eating** each prey, as he was able to show experimentally that this period was linearly related to the weight of the particular prey item, and that the constant involved was a reflection of the feeding rate of the predator in time per unit

Table 39. The relationship between pursuit time (TP) and hunger (after Holling 1966)

1. The speed of the predator in pursuit is a constant, VP.

2. The average distance of pursuit (r_m) for a particular level of hunger can be obtained from the expression for the area of reaction by assuming that it is approximately equal to the radius of a circle of equal area:

   i.e. Given that $A = (KA)\,r_o^2$

   and assuming $\pi\, r_m^2 = (KM)\,r_o^2$

   then $$r_m = r_o\sqrt{\frac{KA}{\pi}}$$

   Defining KR as a constant equal to $\sqrt{\frac{KA}{\pi}}$

   then $$r_m = (KR)\,r_o$$

3. But the predator pursues only until within striking distance (DS) of the prey. Hence pursuit distance (DP) is given by:

   $$DP = r_m - DS$$
   $$= (KR)\,r_o - DS$$

   Substituting for r_o to introduce hunger terms we have:

   $$DP = (KR)(GM)(H - HT) - DS \qquad H > HTP$$
   $$DP = 0 \qquad H \leq HTP$$

4. Given the constant speed of the predator we can then calculate pursuit time (TP):

   $$TP = (KR)(GM)(H - HT) - DS/VP \qquad H > HTP$$
   $$TP = 0 \qquad H \leq HTP$$

Table 40. The relationship between eating time and hunger and time in the digestive pause and hunger (after Holling 1966)

Eating Time

For mantids, eating time (TE) was related only to the weight of the prey item (WE) and was independent of hunger,
whence:

$$TE = KE(WE)$$

where KE is a constant, the time to consume 1gm of prey.

Time in Digestive Pause

We have expressed hunger as:

$$H = HK + (HO - HK)\,e^{-AD(T)}$$

which equation we can rearrange to obtain:

$$T = \frac{1}{AD}\ln\left(\frac{HK - HO}{HK - H}\right)$$

This is the time taken for the hunger level to change from HO (the level immediately after eating) to H. Now the time in digestive pause (TD) is that time required for the hunger level to change from HO to HT, the threshold for attack, therefore:

$$T = \frac{1}{AD}\ln\left(\frac{HK - HO}{HK - HT}\right) \qquad HO > HT$$

and

$$TD = 0 \qquad HO \leq HT$$

weight. The equation he used is shown in the first part of table 40. Holling next turned to the finding of an expression for the time spent in **digestive pause**, and was able to derive this more or less directly from the expression of hunger which includes time as an unspecified variable. This is equation 4 in table 33, which can be rearranged to place the time variable on the left hand side. It is then a simple matter to substitute HT (that is the hunger level at the threshold of attack) for the unspecified hunger level used in the earlier form. The left hand side of the equation is then an expression of the time in digestive pause, that is the time for the hunger level to rise from the low point it reached immediately following the particular predation event to the threshold level for further attack.

Holling now had a series of expressions for all of the subcomponents identified in his initial analysis of the predation process. Some of these are of a general form, others are strictly empirical, relating solely to the particular system he studied in the laboratory. These so called fragmental equations were then used as building blocks in the synthetic model that he put together in order to simulate the whole predation process. This synthesis required modification of some of the expressions used and followed the pattern outlined below.

Holling argued that during any period of daylight a particular predator, the mantids in this case, would go through a series of attack cycles, each of which could be considered to have four

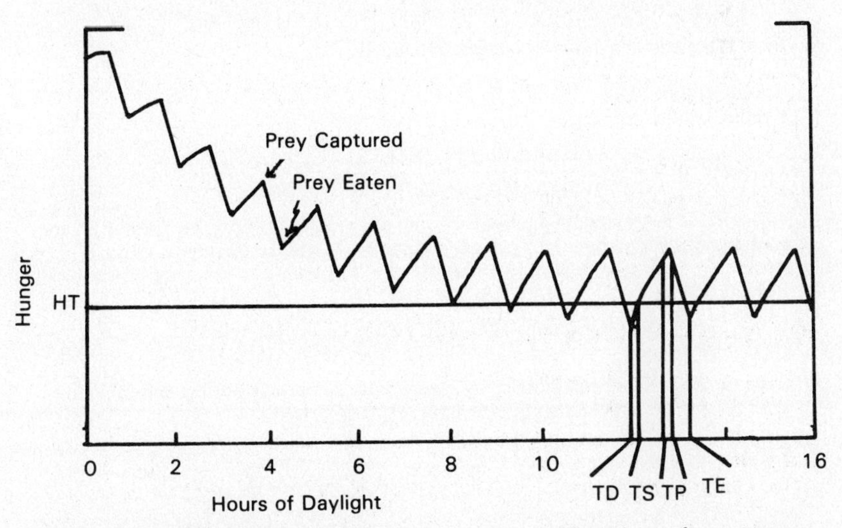

Fig. 51 The daily pattern of feeding cycles used in Holling's invertebrate response model (after Holling 1966)

parts: the time in digestive pause, the searching time, the pursuit time and the eating time. He envisaged these interacting with the hunger level at any point during the day in the fashions shown in figure 51. Holling was then able to construct his simulation model from the fragmental equations by converting them into expressions for each of these temporal segments. This required the introduction of one additional variable, that Holling called H1, which was the level of hunger at any point in time during the day, the current state of hunger as it were. He then rewrote some of the fragmental equations to include this variable where appropriate. Table 41 shows the equations he used in the synthesis. It will be observed that the expressions for time spent in digestive pause, pursuit time and eating time are only minor modifications of the versions developed earlier when the particular processes were being studied piecemeal. His

Table 41. Synthesis of equations to simulate the attack cycle of a predator (after Holling 1966)

1. **The temporal structure of the cycle**

$$TI = TD + TS + TP + TE$$

where TI is the total time for the cycle, TD is the time spent in digestive pause, TS is search time, TP is pursuit time and TE is eating time.

2. **Time spent in digestive pause (TD)**

$$TD = \frac{1}{AD} \ln\left(\frac{HK - HO}{HK - HT}\right) \qquad HO > HT$$

$$TD = 0 \qquad HO \leq HT$$

from fragmental equation developed in table 40.

3. **Searching time (TS)**
Simple version, see text.

$$TS = \frac{[1 - \pi (RD)^2 (SR) (SP) (SS) NO]}{2 (VR) (RD) (SR) (SP) (SS) NO}$$

where RD is the average distance of reaction of predator for prey, SR, SP and SS are success rates for recognition, pursuit and strike, NO is the density of prey.

4. **Pursuit time (TP)**

$$TP = (KR) (GM) (H1 - HT) - DS/VP \qquad H1 > HTP$$

$$TP = 0 \qquad H1 \leq HTP$$

from the fragmented equations developed in table 39 but substituting **current** hunger level H1 for the unspecified hunger term, H.

5. **Eating time (TE)**

$$TE = KE (W) \qquad W \leq H1$$

$$TE = KE (H1) \qquad W > H1$$

where W is the actual weight of a particular prey item.

expression for **searching time**, however, required a more complex derivation. He approached this derivation by considering the area of the encounter path swept by a predator as it moves during the day, in a fashion reminiscent of his treatment of the prey velocity equation given in table 37. In the simpler case he assumed that hunger level would not change during this searching process and the area swept by the predator was then simply computed as a rectangle plus two semicircles. This could be combined with the prey density and the expressions for recognition, pursuit and strike success to obtain the expression shown in step 3 of table 41. In developing this model, Holling was aware of the fact that the hunger level of the predator would change during the searching process, and the shape of the area swept by the predator would change accordingly. To take account of this he derived a much more complex expression for searching time, which took into account the irregular shaped area that would be swept by a predator whose hunger level was increasing in a specified manner. I have not included this complex derivation here, as its detailed form is not necessary information for a working understanding of the invertebrate response model. Full details of Holling's procedure are given in the text and appendices of his 1966 paper.

Having obtained these four equations relating the four segments of the attack cycle to the hunger level of the predator, Holling synthesized these into a simulation model using a logic similar to that laid out in figure 52. As with previous models considered, it was necessary to provide initial values for the level of hunger and to read in values for the various constants discussed earlier. In addition, a book-keeping process was necessary to keep track of the passage of time, so that the end of the daily feeding period could be taken into account in simulating the day's events.

In an attempt to validate the model, Holling carried out a series of further laboratory experiments in which he exposed flies at a variety of densities to the depredations of a mantid throughout a daily feeding cycle. He used similar densities and time periods in a run of the simulation model, and compared the rates of attack predicted by the model with the actual rates observed in his experimental situation. Figures 53 and 54 show some of these results and the comparison Holling made between them and the predictions of the model. It is evident from these figures that there was substantial agreement between the predictions and the data obtained in the experimental situations.

Holling concludes his account of the invertebrate response

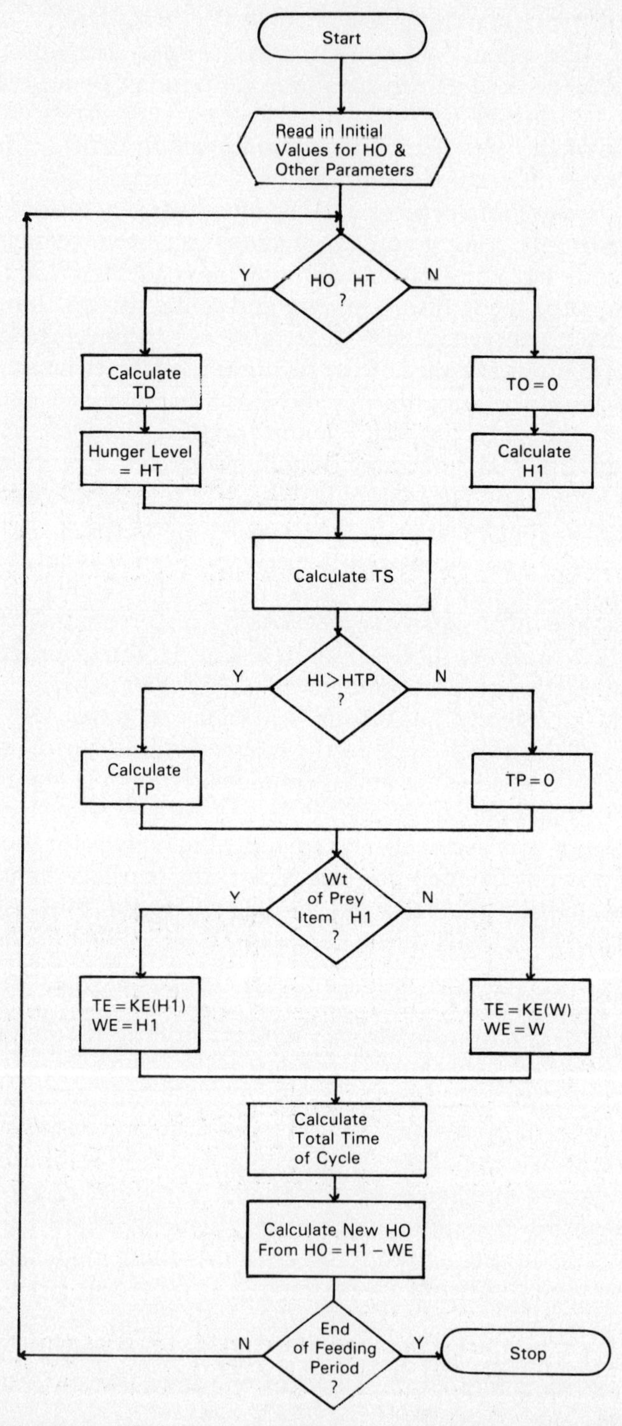

Fig. 52 Flowchart of the invertebrate response model of Holling (after Holling, 1966)

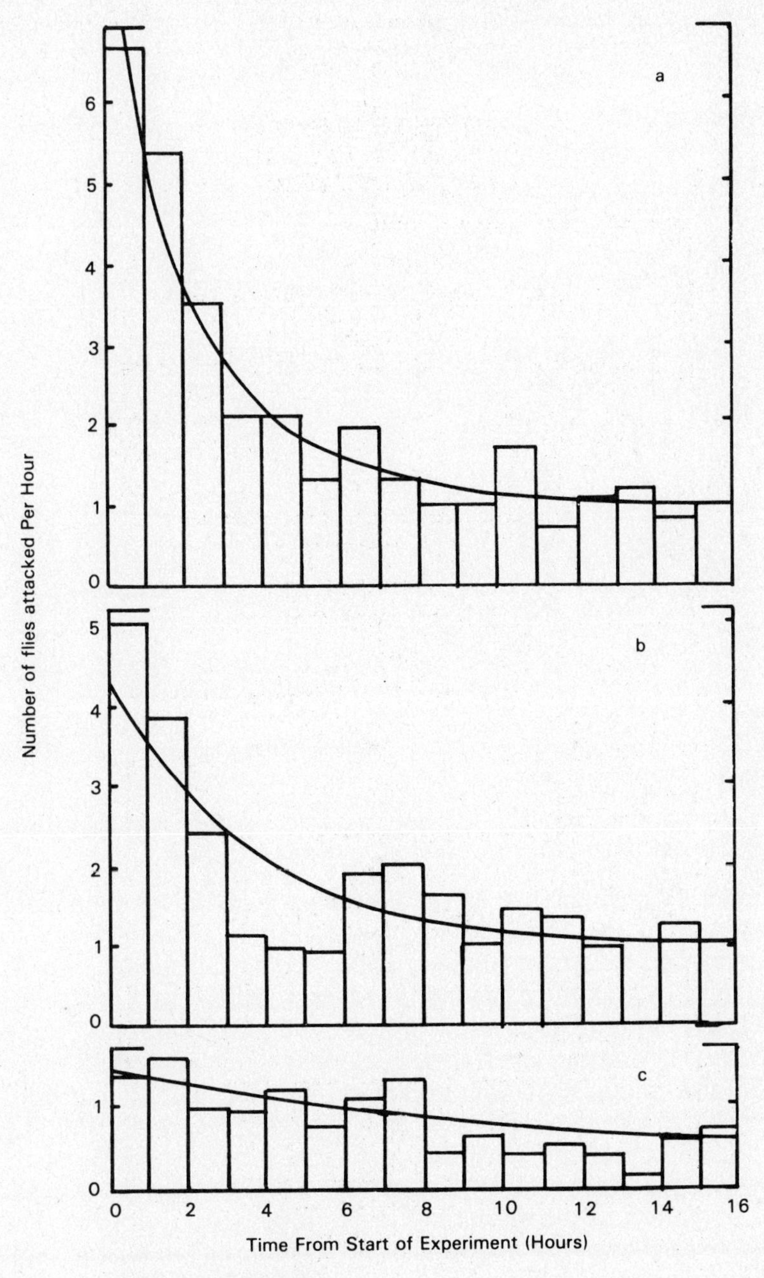

Fig. 53 Comparisons of the predicted and observed rates of predation of the mantid, *Hierodula crassa* with prey at a. high, b. intermediate and c. low prey densities. The predictions were obtained using Holling's invertebrate response model (after Holling 1966)

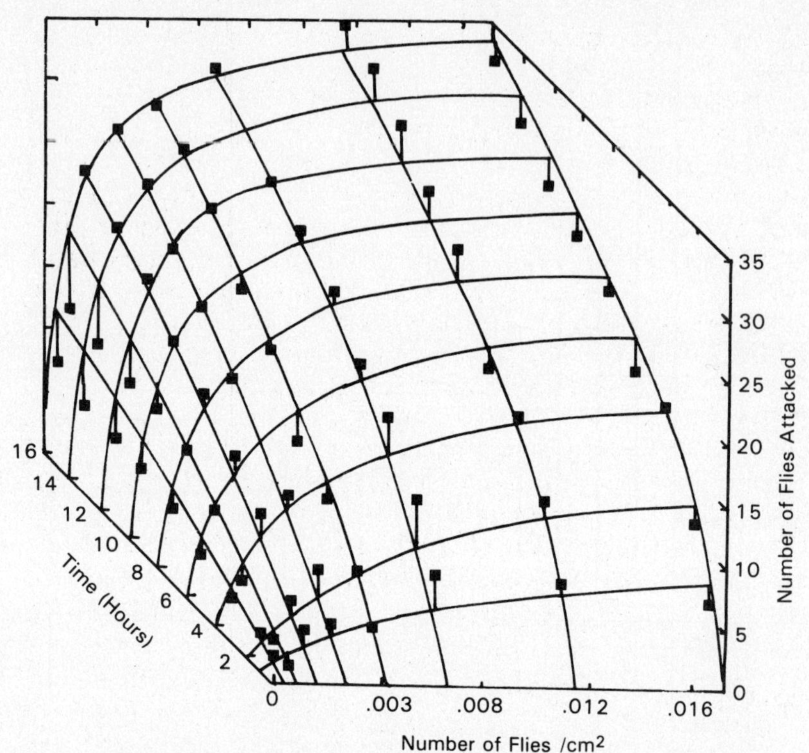

Fig. 54 The response surface of Holling's invertebrate response model relating the number of flies attacked by a mantid to fly density to time. The points are experimental data which can be compared with the predicted surface (after Holling 1966)

model by a discussion of the perceived goals of his modelling efforts and the extent to which these goals are achieved. He comments elsewhere that a good ecological model should attempt to maximize **generality, realism, holism** and **precision.** He suggests that his approach, expressed in the sort of work he did in deriving the invertebrate response model and other more complex models of the predation process, comes close to achieving this aim, although it is generally accepted that the goals of realism and generality, precision and holism are to some extent opposing pairs. Holling reviewed a large number of cases of predation in the discussion of his 1966 paper, showing that the invertebrate response model was appropriate in a large class of cases given suitable values for the various constants and other parameters in the model. He concluded, however, that where learning was a significant part of the behaviour of the animal then the invertebrate response model would fall short. He incorporated some of the necessary additional information into his next model, the 'vertebrate response' model, published in 1965.

There can be little doubt that Holling's work in general has given us many new insights into the process of predation, as well as establishing a bench mark in methodology and philosophy that can be applied to a variety of ecological studies. His models are rigorous, based on thorough experimentation in the laboratory and validated against independent laboratory data. An examination of the literature suggests that they have a wide applicability and have been used in part in several applications, particularly in building models of pest management situations. Following the publication and widespread dissemination of Holling's work however, considerable criticism was levelled against it for its failure to include any field verification of the models produced, or indeed any evidence at all that the experimental components approach could be applied to the more complex field situation. It was left to Frazer and Gilbert some ten years later, in 1976, to make the jump between the laboratory based models such as those described by Holling and a version based on the same philosophy but applied to a field situation. These authors studied the predatory relationship between ladybirds and aphids in laboratory and field; they followed a sequence of activities very similar to that carried out by Holling in the laboratory to obtain a model of the predation process operating in this system. However they then turned to a study of a parallel system in the field. Their major conclusion was that, with respect to the study of the predation process itself, the models derived in the laboratory were so detailed that they could be applied only in the controlled circumstances possible in laboratory situations. Frazer and Gilbert showed, however, that the general approach through components analysis of an ecological process could be applied in field situations; they built a simpler model along these lines including the predator and prey densities, predator voracity, prey age distribution and temperature, which fitted their field data admirably. They point out that temperature has an overriding effect in the field situation and is an essential component both within the predation process and on population dynamics in the field. They note that temperature has a differential effect on population dynamic processes, operating through rates of development, and on the predation process itself, operating through effects upon rates of activity. The concluding statement of these authors may be taken as a fitting conclusion to our account of Holling's work: "(1) laboratory studies of ecological relationships must not be trusted until verified in the field and (2) it is in fact possible to make detailed predator-prey studies in the field, to explain the observed impact of predation on the prey population''. Such detailed and elegant predator-prey studies have been made, subsequently, by Frazer, Gilbert and their colleagues in British Columbia and are presented in the compendium by Baumgartner *et al.* (1981).

7.2.2 Movement

The experimental components approach pioneered by Holling provides a powerful philosphy for the study of any ecological process. In some of his more reflective papers, Holling (1964, 1965) identified particular processes that he thought would repay study in this fashion. One was the process of **animal movement**, and it was with this in mind that I undertook studies of the movement process using the sort of approach identified by Holling. The resulting model is by no means as far advanced as Holling's study of the predation process, but it does provide a basis for further study, and indeed such study is in progress at the time of writing. A full account of the first model I built is given in a 1971(a) paper which will be paraphrased here to provide an account appropriate to the purposes of this work.

Studies made of the populations of insects occurring in water-filled tree-holes in England (Kitching 1971b and 8.1) led to the realization that for many organisms the process of habitat to habitat movement was crucial for their survival from generation to generation. Such **discrete habitats** are widespread in ecosystems and include such well-known sites as carrion, dung pats, annual plants, stored products, nests, fungus fruiting bodies and so on. The model of movement I built was conceived of in these terms and was part of an on-going study of the relationships between animal population dynamics and habitat characteristics. Although I initially thought of them as **habitats**, I now refer to the fixed points in the movement space as **sinks**, to indicate that the processes and mechanics involved are common to any set of resources dispersed in a spatially heterogeneous fashion. The wider body of ecological theory into which this work feeds is the subject of papers such as that of Southwood (1977) and Kitching (1977).

To explain the modelling effort in the present context, it is necessary to relate the features of this process to those identified in the account of predation. One can identify numerical and

Table 42. The structure of movement

1. Biological component effects
2. Environmental component effects
3. Numerical effects
These correspond to Holling's (1961) functional (1 and 2) and numerical (3) responses within the predation process.

Table 43. Biological components of the movement process

Departure Phase
Fuel supply
Ability to travel
Social facilitation
Starting conditions
Transit Phase
Directionality
Speed
In-transit mortality
Persistence
Arrival Phase
Detection
Exploitation

functional responses with respect to the movement process, and in similar fashion one can draw up a series of components reflecting the structure of the movement process. Tables 42 and 43 show one approach to this process of component identification. As with the predation process, the components approach to movement is concerned primarily with the functional responses involved in movement from place to place and considers, therefore, the **biological** and **environmental components** of movement which bear upon these relationships. Of the three temporal phases identified in the movement process, the **departure, transit** and **arrival phases,** the model concentrated largely upon the **transit** phase. It was an extremely simple model and in many ways analogous, in terms of its level of complexity, with the disc equation model of predation, rather than any of Holling's later more complex attempts. The components chosen for inclusion in the version of the model presented here were the **speed** and **directionality** of movement of the organisms concerned, the **in-transit rate of mortality** and the **distribution and attractiveness of sinks** in the environment of the moving organisms.

The basic idea behind the model was to construct a two-dimensional grid with a number of sinks scattered in it and to start a known number of organisms dispersing from a central point. Using a simulation approach, I aimed to record the numbers of organisms reaching each sink and, using these results, to obtain predictions and generate further hypotheses about the effects of distance of sinks from the starting point, patterns of movement, and mortality rates on the success of moving organisms, measured in terms of the numbers that arrived at sinks during the particular movement event. As we have observed in our examination of other models, a number of decisions about the construction and level of precision of the model have to be made before proceeding to the simulation phase, and these decisions may be summarized in three groups.

1. The first set of decisions related to the **pattern of sinks** to be set up within the chosen environmental grid. Throughout the model, the aim was to be as general as possible; therefore this decision, and several later ones, were intended to maximize this **generality.** The habitats were assumed to be distributed in a uniformly random manner within the grid. This was achieved in the model by generating x and y coordinates for the number of sinks required, using the uniform random number generator in the computer. In addition, each of the sinks was assumed to have around it a circle of influence whose radius would be directly proportional to the degree of attractiveness of the particular site. In the real world, this would reflect the stage, size, age or some other key property of the sink, be it a flower, potential oviposition site or what have you. The way in which the radii of influence are distributed among the sinks will be a characteristic of the particular type of sink involved and the properties of the sensory apparatus of the moving animal; in order to maintain generality, a normal distribution was used for these radii between two limits assigned arbitrarily so as to keep the average size of the circles of influence within bounds defined by the total size of the grid used. A particular radius was obtained by selecting a value using a computer subroutine which generated random normal deviates (see chapter 5) and scaling the number obtained within the chosen limits. The further assumption was made that, once a moving organism crossed the boundary of the circle of influence of a particular sink, it could be said to have achieved its goal and hence could be removed from any further consideration in the simulation.
2. The second group of decisions concerned the actual **movement patterns** of the organisms. In the interest of

generality, the decisions were based on observations made of arthropods moving in a near uniform environment. These suggested that the pattern of movement of an animal in two dimensions could be mimicked using a series of frequency distributions. The first of these related to the **starting directions** adopted by organisms released at a central point. It was shown from laboratory studies at least, that there was an equal probability of a particular organism setting off in any direction in the 360 degrees available. Therefore, in terms of the model, it was a straightforward matter to assign a starting direction using a uniform random number generator. The second distribution reflected **subsequent directions** of movement. It was considered that an animal moving in a particular direction in one time interval was most likely to continue moving in the same direction in the next time interval, with probabilities decreasing for increasing deviations from the straight and narrow track. The pattern of deviation from previous direction is adequately represented by a normal distribution, with the mode of the distribution representing the deviation class around zero degrees. Such an assumption is supported by analyses of tracks measured in laboratory arenas for a variety of animals, including moths, beetles and snails. From a modelling point of view, there is considerable advantage in adopting such a normal distribution for the purposes of choosing directions of movement in any time interval, since a generator of random normal deviates can be used again and the standard deviation of the normal distribution, which must be specified in the use of such a generator, can be employed as an easy way of building-in any given degree of directionality. If a very small standard deviation is chosen, it suggests a very compact and "pointed" distribution, with a very high probability of the direction chosen being the same as or very close to the previous direction. On the other hand, if the specified standard deviation is large, there will be a much greater likelihood of major departure from the previous direction when this quantity is chosen from the specified normal distribution. Lastly, it was necessary to specify the nature of the distance moved in any unit of time. In fact, in the model described in 1971, constant values for displacement per unit time were entered for any particular run of the model. In subsequent versions this was replaced by a stochastic process, where the direction to be moved in any time interval was chosen from a normal distribution of specified mean and standard deviation. The actual results of the simulations differed little from those presented here, which employed the mean value only.

3. The last set of decisions concerned the **mortality process** to be imposed on organisms moving between sinks and outside the circles of influence involved. Again with generality in mind, it was assumed that the mortality per unit time would be a constant percentage of the animals surviving up to the start of any time period. This of course, was an oversimplification, but could be made more realistic if a specific system were being studied. If we assume that the environment between sinks is more or less homogenous and that the probability of death for a particular organism is independent of the age of that organism (that is to say a "Type 2" survivorship curve, see, for example, Krebs 1972), then the mechanism as applied in the model would be close to a realistic one.

The basic structure of the model, which synthesized these decisions into a representation of the movement process, is depicted as a flowchart in figure 55. As in most ecological models, the basic procedure involved was a large time loop, preceded by various initialization procedures and followed by a book-keeping and output process. Within the time loop of the movement model, two separate submodels can be identified. The first of these was called the **movement submodel**, this being the process by which the surviving organisms at that point in time were rearranged in space, and the second, the **mortality submodel**; the processes by which a certain number of the surviving organisms were deemed to have died in that time interval. The x and y coordinates of the sinks and their associated radii of influence were generated using a separate programme, and were read in before entry into the time loop, together with the chosen mortality rate, the variance of the normal distribution to be used in selection of directions, the displacement per unit time and the initial number of organisms.

The **movement submodel** is diagrammed in figure 56. The first process was to take the known (x,y) coordinates of the organism (0,0 initially for all organisms) and to compute a new set of such coordinates using the rules already mentioned. Thus, if the move considered was the first in the simulation run, then directions where chosen at random and new coordinates calculated knowing these directions and the average distance of

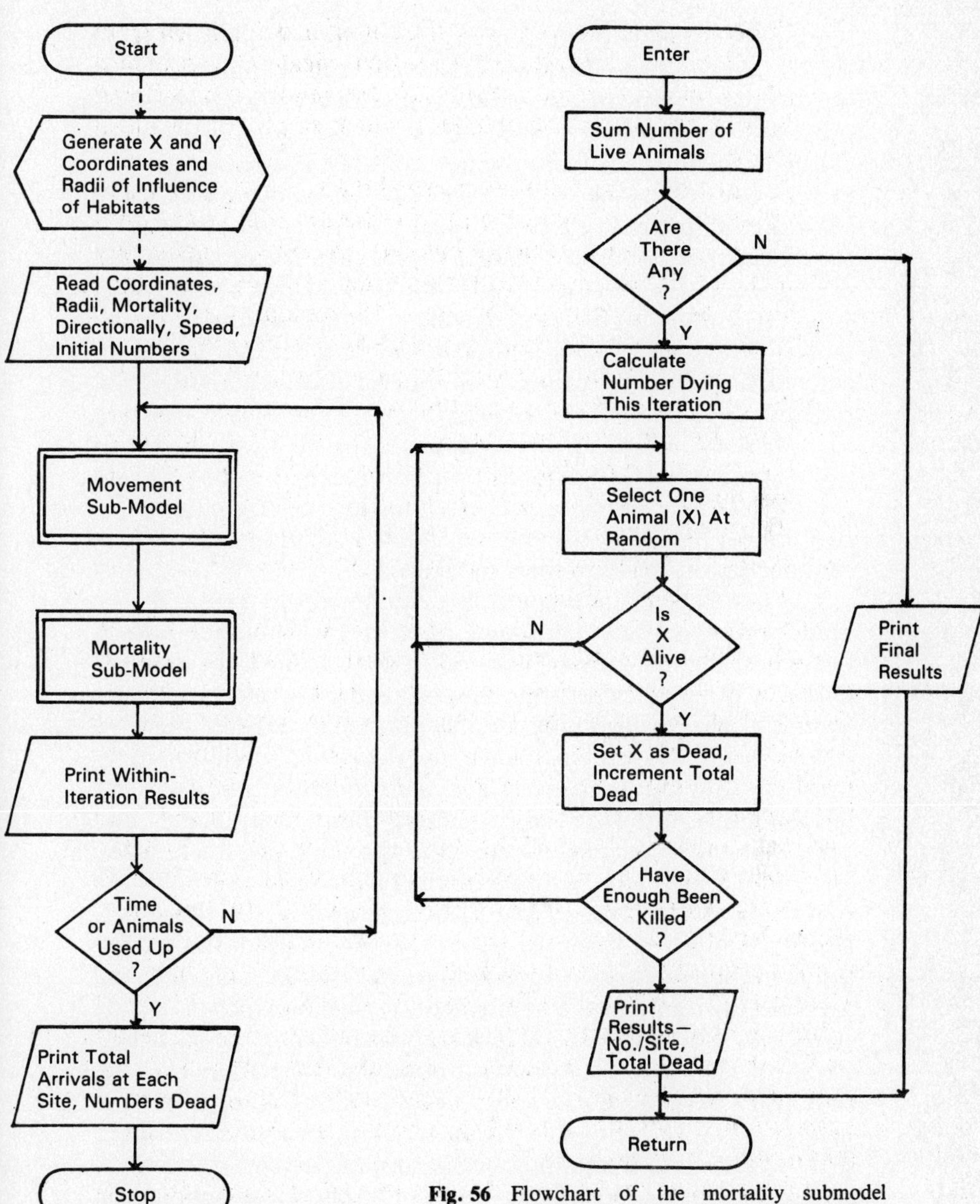

Fig. 55 Overall flowchart of the component model of movement among units of discrete habitats (after Kitching 1971a)

Fig. 56 Flowchart of the mortality submodel within the component model of movement (after Kitching 1971a)

displacement per unit time. In all time intervals other than the first, a similar procedure was involved, except that directions were chosen from the specified normal distribution. The next process was to consider each surviving organism in turn, to calculate the distance from the new position of the organism to each of the sinks using simple trigonometry, and to compare the calculated distance in each case with the radius of influence. If the distance was shorter than the radius then the organism was deemed to have arrived at the habitat involved and was removed from the simulation thereafter. A slight complication not shown in the flowchart covered those instances where an organism entered the circle of influence of two adjacent habitats within the same time interval. In these circumstances a simple arbitrary rule was used, assigning the organism to the site whose radius of influence had been penetrated proportionately farthest. Having completed the movement process, the flow of computation in the model moved to the mortality submodel which is summarized in the flowchart in figure 57.

The first process in the **mortality submodel** was to sum up the number of organisms remaining alive in the simulation and, if this number was greater than zero, to calculate the number dying in the particular time interval, using the mortality rates specified at the beginning of the particular run. In the case where no animals remained alive, the flow of computation moved next to the output process and the simulation run itself was terminated, following appropriate data summation and printing-out. Returning to the mortality process where there were animals left, the required number were selected from those remaining alive in the simple fashion shown in the flowchart, and were removed from the pool of survivors accordingly. The last procedure within the time loop was a book-keeping one, and involved printing-out the numbers of organisms which arrived at particular sites and the total number which died in the particular iteration. The flow continued around the time loop until no organisms were left or the assigned computing time expired. The simulation was completed by a totalling procedure and the printing-out of the total number of animals arrived at each site and the total numbers dying during the particular simulation. Output of the type shown in table 44 was obtained and, in addition, pictorial output of the sort shown in figure 58 was possible. This graphical mode of output used a plotter of the type referred to in chapter 4, and was a useful heuristic tool in the interpretation and presentation of the results of the modelling process. Two such plots are shown in the figure and

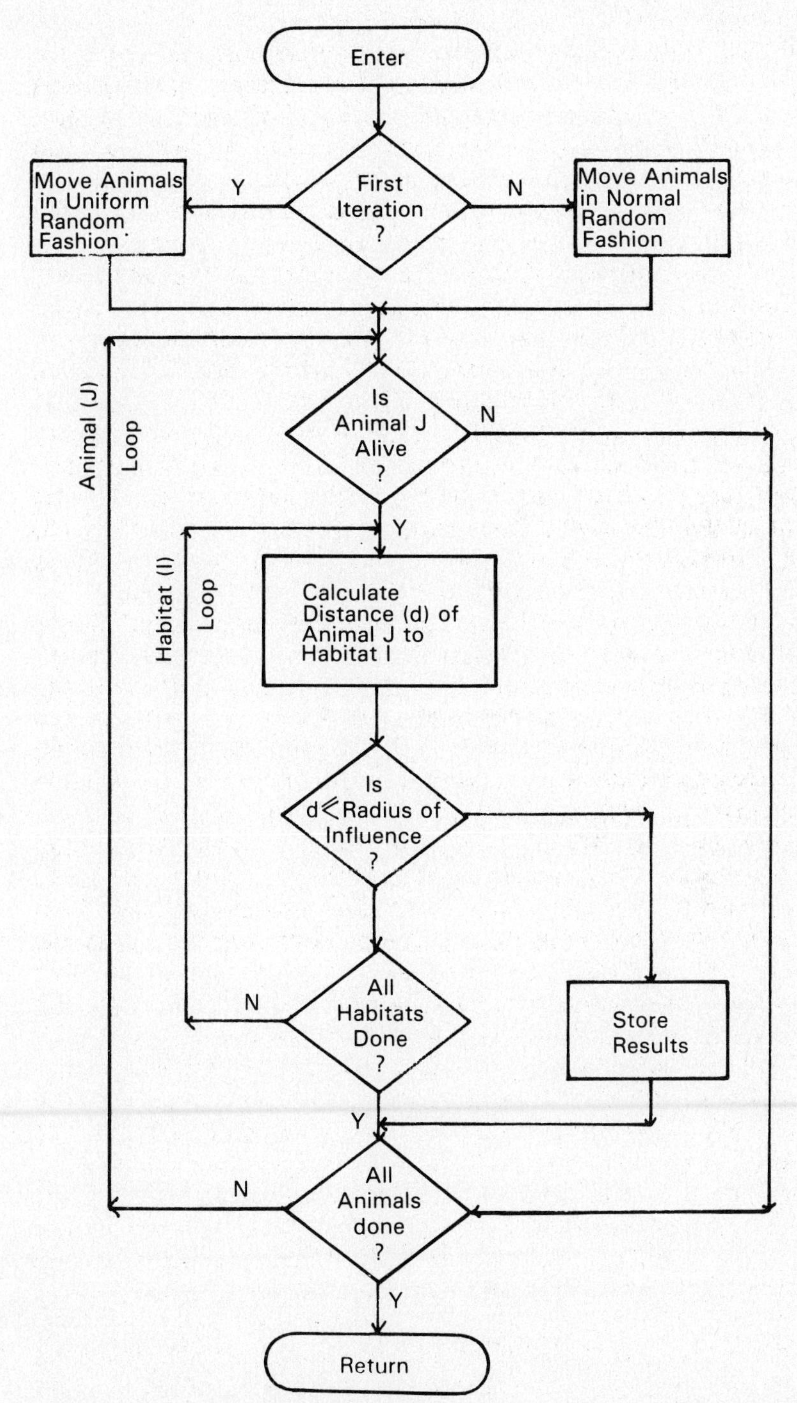

Fig. 57 Flowchart of the movement submodel within the component model of movement (after Kitching 1971a)

Table 44. Example of print-out of results from model of Kitching (1971a)

Iteration	Numbers of animals reaching each site																									Number
	1	2	3	4	5	6	7	8	9	10	11	12	13	14	15	16	17	18	19	20	21	22	23	24	25	dying*
1	—	—	—	—	—	—	—	—	—	—	—	—	—	—	—	—	—	—	—	—	—	—	—	—	—	3
2	—	—	—	—	—	—	—	—	—	—	—	—	—	—	—	—	—	—	—	—	—	—	—	—	—	3
3	—	—	—	—	—	—	—	—	—	—	—	—	—	—	—	—	—	—	—	—	—	—	—	—	—	3
4	—	—	—	—	—	—	—	—	—	—	—	—	—	—	—	—	—	—	—	—	—	—	—	—	—	3
5	—	—	—	—	—	—	—	—	—	—	—	—	—	—	—	—	—	—	—	—	—	—	—	—	—	3
6	—	—	—	—	—	—	—	—	5	—	—	—	—	—	—	—	—	—	—	—	—	—	—	—	—	2
7	—	—	—	—	—	—	—	—	6	—	—	—	—	—	—	—	—	—	—	—	—	—	—	—	—	2
8	—	—	—	—	—	—	—	—	2	—	—	—	—	—	—	—	—	—	4	—	—	—	—	—	—	2
9	—	—	—	—	—	—	—	—	—	—	—	—	—	—	—	—	—	—	1	—	—	—	—	—	—	2
10	—	—	—	—	—	—	—	—	1	—	—	—	—	—	—	—	—	—	—	2	—	—	—	—	—	2
11	—	—	—	—	—	—	—	—	—	—	—	—	—	—	—	—	—	—	1	—	—	—	—	—	—	2
12	—	—	—	—	—	—	—	—	—	—	—	—	—	—	—	—	—	—	1	—	—	—	—	—	—	2
13	—	—	—	—	—	—	—	—	1	—	—	—	—	—	—	—	—	—	2	1	—	—	—	—	—	1
14	—	1	—	—	—	—	—	—	1	—	—	—	—	—	—	—	—	—	—	2	—	—	—	—	—	1
15	—	—	—	—	—	—	—	—	—	—	—	—	—	—	—	—	—	—	—	—	—	—	—	—	—	1
16	—	—	—	—	—	—	—	—	—	—	—	—	—	—	—	—	—	—	—	—	—	—	—	—	—	1
17	—	—	—	—	—	—	—	—	—	—	—	—	—	—	—	—	—	—	—	—	—	—	—	—	—	1
18	—	—	—	—	—	—	—	—	—	—	—	—	—	—	1	—	—	—	—	—	—	—	—	—	—	1
19	—	—	—	—	—	—	—	—	—	—	—	—	—	—	—	—	—	—	—	—	—	—	—	—	—	1
20	—	—	—	—	—	—	—	—	—	—	—	—	—	—	—	1	—	—	—	—	—	—	—	—	—	1
21	—	—	—	—	—	—	—	—	1	—	—	—	—	—	—	1	—	—	—	—	—	—	—	—	—	1
22	—	—	—	—	—	—	—	—	—	—	—	—	—	—	—	—	—	—	1	—	—	—	—	—	—	1
23	—	1	—	—	—	—	—	—	—	—	—	—	—	—	—	—	—	—	—	—	—	—	—	—	—	1
24	—	—	—	—	—	—	—	—	1	—	—	—	—	—	—	—	—	—	—	—	—	—	—	—	—	1
25	—	—	—	—	—	—	—	—	—	—	—	—	—	—	—	—	—	1	—	—	—	—	—	—	—	1
26	—	—	—	—	—	—	—	—	—	—	—	—	—	—	—	—	—	—	—	—	—	—	—	—	—	1
27	—	—	—	—	—	—	—	—	—	—	—	—	—	—	—	—	—	—	—	—	—	—	—	—	—	1
28	—	—	—	—	—	—	—	—	—	—	—	—	—	—	—	—	—	—	—	—	—	—	—	—	—	1
29	—	—	—	—	—	—	—	—	—	—	—	—	—	—	—	—	—	—	—	—	—	—	—	—	—	1
30	—	—	—	—	—	—	—	—	—	—	1	—	—	—	—	—	—	—	—	—	—	—	—	—	—	1
31	—	—	—	—	—	—	—	—	—	—	—	—	—	—	—	—	—	—	—	—	—	1	—	—	—	1
32	—	—	—	—	—	—	—	—	—	—	—	—	—	—	—	—	—	—	—	—	1	—	—	—	—	1
33	—	—	—	—	—	—	—	—	—	—	—	—	—	—	—	—	—	—	—	—	—	—	—	—	—	1
34	—	—	—	—	—	—	—	—	—	—	—	—	—	—	—	—	—	—	—	—	—	—	—	—	—	1
35	—	1	—	—	—	—	—	—	—	—	—	—	—	—	—	—	—	—	1	—	—	—	—	—	—	1
36	—	—	—	—	—	—	—	—	—	—	—	—	—	—	—	—	—	—	—	—	—	—	—	—	—	1
37	—	—	—	—	—	—	—	—	—	—	—	—	—	—	—	—	—	—	—	—	—	—	—	—	—	1
38	—	—	—	—	—	—	—	—	—	—	—	—	—	—	—	—	—	—	—	—	—	—	—	—	—	1
39	—	—	—	—	1	—	—	—	—	—	—	—	—	—	—	—	—	—	—	—	—	—	—	—	—	0
Totals	0	3	0	0	1	0	0	0	18	0	1	0	0	0	1	2	0	1	11	5	1	1	0	0	0	55

*These figures are rounded to the integer value above the calculated 3 per cent per iteration value.

Note: This example used a starting number of 100, a mortality rate of 3 per cent per iteration, a variance of the normal distribution of 0.75, and an increment/iteration of 0.20

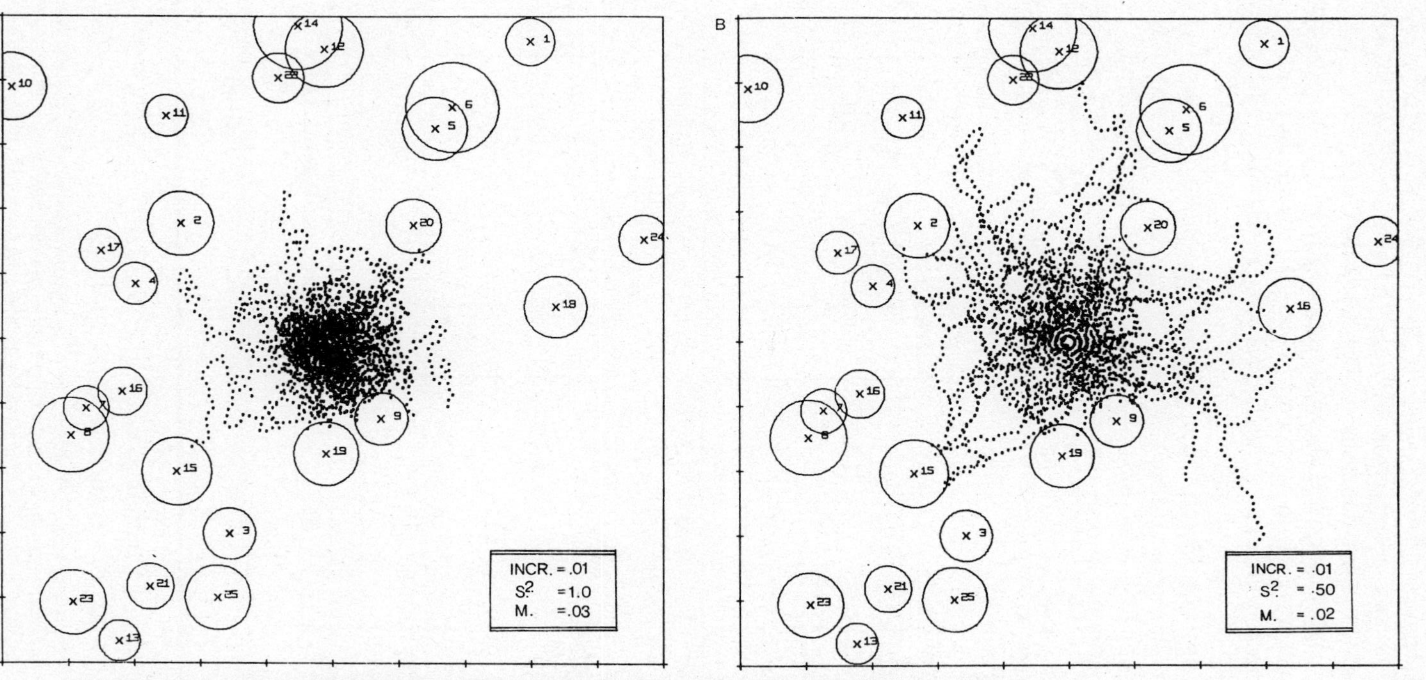

Fig. 58 Scatter diagrams generated by the component model of movement under various conditions
Abbreviations: INCR increment per iteration; S^2 variance of the normal distribution; *M* percentage mortality per iteration (from Kitching 1971a)

permit instant comparison between the results of two different simulations using the same model.

The model described was investigated in some detail and comparisions were made between the output generated by the model and a variety of data gleaned from the literature. Predictions of three kinds were obtained.

1. The first related the number of organisms arriving at particular sinks to the distance of that sink from the starting point of the movement event. Figure 59 shows the predicted curves denoting this relationship under a variety

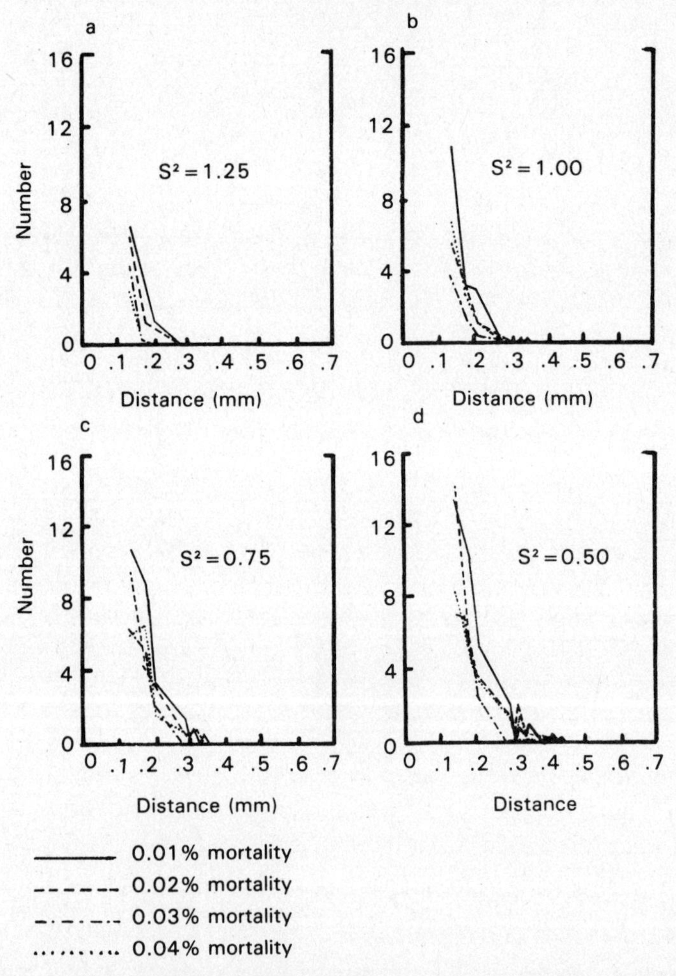

Fig. 59 Predicted number of animals arriving at particular habitats, plotted against distance from starting points for four levels of directionality (S_2) and four levels of mortality (after Kitching 1971a)

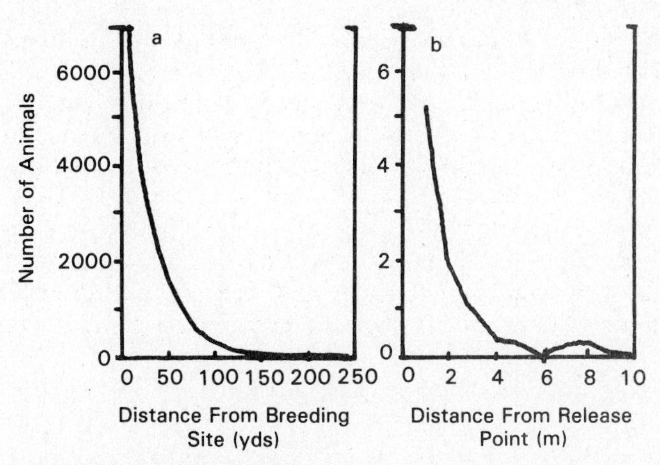

Fig. 60 Curves of number versus distance from starting point for *Culicoides impunctatus* (data from Kettle 1951) and *Armadillidium vulgare* (data from Paris 1965) (after Kitching 1971a)

of starting conditions for the model. This sort of output from the model was obtained to permit comparison with the results of a great many field and laboratory experiments which were available in the literature. Figure 60 shows two of these sets of results from other studies, and shows a shape and functional form for the actual result which corresponds closely with that predicted by the model.

2. The second kind of output obtained by runs of the model is summarized in figure 61. This shows the relationship between percentage success — that is, the proportion of the organisms which reached sites during a particular movement event — plotted against the imposed mortality rate on one axis and the variance of the normal distribution used to simulate different degress of directionality on the second axis. These predictions were unverified at the time of publication.
3. The third prediction obtained by the model was the relationship between the percentage success, in the sense already used, and the displacement per unit time used to denote the speed of movement in any particular run. A strictly linear relationship between these two variables was obtained and a sample result is given in figure 62. An examination of the natural history of a variety of insects inhabiting water-filled tree-holes, the study of which had led to the construction of the model under discussion, pro-

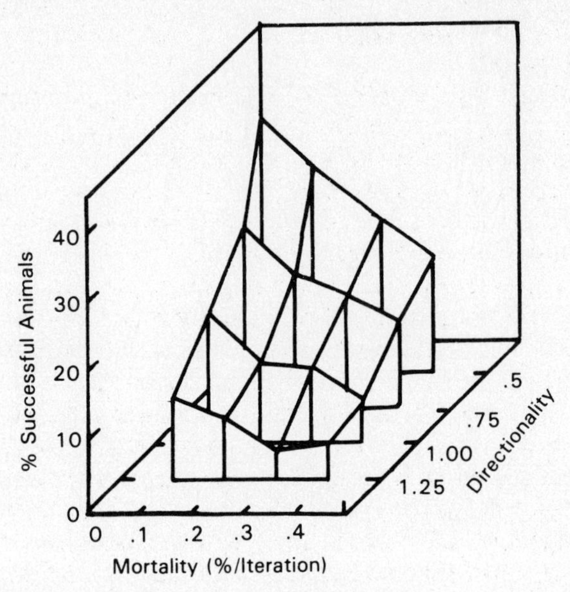

Fig. 61 Predicted graph of percentage of successful dispersers versus mortality rates versus directionality (after Kitching 1971a)

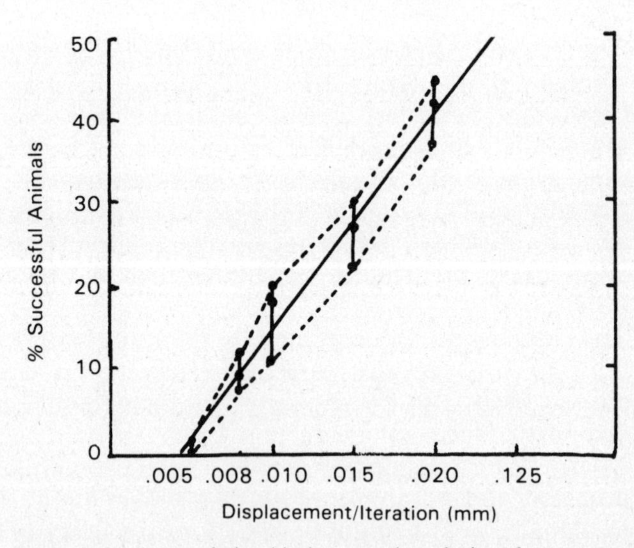

Fig. 62 Graph of the linear relationship between the velocity of movement and the percentage success as predicted by the component model of movement (after Kitching 1971)

vided anecdotal support for the relationship predicted by the model.

Overall the insights provided by the model into the process of movement in a heterogeneous environment, that is one in which there are a number of sinks which are more or less attractive to the moving organisms, were sufficient to justify its construction. These included an understanding of the various options open to an organism in evolutionary terms, when a movement event of the kind simulated is an integral part of its life history. For example, it was apparent from the simulations that an organism could achieve a high probability of arrival at one of the sinks, either by maximizing its directionality of movement or by increasing its speed of movement. Intermediate situations, of course, would be the most common, where the degree of directionality was balanced off against the mean speed of movement. Analysis of the results showed that the model was very sensitive to the value of the in-transit mortality rate and small changes in this rate could have highly significant effects on the outcome of the movement event. The model allows identification of a great many areas where our knowledge is less than complete and from it a whole scheme of experimental work for the further investigation of the process of movement can be and has been devised. Subsequent work, building upon this approach, has been carried out and the interested reader is referred to Kitching and Zalucki (1982) and Zalucki and Kitching (1982).

7.2.3 Prospects in Component Modelling

Consideration of the component models of predation, and to a lesser extent those of movement, shows that the primary purpose of the model is descriptive. This is not a primary function in the population models in chapter 6 nor in the models of communities to be examined in chapter 8. The aim is to analyze the process, break it down into its component parts, investigate each of these component parts separately, and then resynthesize these component parts into a general, holistic model of the process as it is observed in a large number of cases in nature. The aim of the work is not primarily predictive, but to obtain as complete an understanding as possible of the mechanics and functional features of the process itself. This is not to say, of course, that this sort of model cannot be used for prediction in management and other applications but this is a secondary spinoff from the modelling efforts.

Table 45. A tentative identification of the ecological components of the reproductive process

1. **The Premating Phase**
   - Level of maturity
   - Environmental influence on endocrine status
   - Social facilitation

2. **The Mating Phase**
   - Sex ratio
   - Densities of the sexes
   - Availability of mating sites
   - Interference
   - Social facilitation
   - Learning
   - Mating history
   - Epideictic feedback

3. **The Post-Mating, Pre-Gravid Phase**
   - Developmental rate
   - Energetic status
   - Effects on activity
   - Abortion

4. **The Parturition Phase**
   - Availability of parturition sites
   - Social facilitation
   - Miscarriage
   - Brood size

This chapter has shown how this approach is applied to the processes of predation and movement, and has referred to work that has been carried out on the competition aspects of the predation process (Griffiths and Holling 1969). Similar work has been carried out on the plant ecological process of photosynthesis (Vinberg and Anisimov 1967). One universal ecological process comes to mind which has not been tackled in this fashion and this is the process of reproduction. It is possible without too much effort to erect schemes of components which could be used to represent the reproductive process. Table 45 is an attempt at this compartmentalization, but it would require a similar sequence of simulation activity, experimental observation and field validation before we could claim to understand the ecological process of reproduction in anything like the fashion that we understand the predation process, and may come to understand the movement process.

The last two chapters have examined the modelling approaches taken to the study of populations and the particular processes acting within and among these populations. The synthesis of these types of models is required if one is to approach the modelling of communities and ecosystems, that is, ecological entities made up of sets of populations which interact in a variety of ways through the general processes already identified. The examples in the next chapter review some of the approaches to this higher level of ecological modelling.

8

Communities and Ecosystems

Chapters 6 and 7 dealt with organisms either grouped into populations or considered as individuals. Much ecological theory has been developed at these levels of integration, particularly associated with ideas in population ecology. Models of systems formulated at these levels deal with flows of organisms as individuals within populations, or as parcels of energy. As intimated in chapter 1, however, the science of ecology also concerns itself with higher levels of integration, indeed, the notion of **community** is one of the central tenets of descriptive ecology. The term **ecosystem**, used initially by Tansley (1935), can be interpreted as a very farsighted vision of the systems approach to ecology, not elaborated until the last decade or so.

The methods and philosophy of systems ecology are well applied to the community and ecosystem as well as the yet higher levels of **biome** and **biosphere.** This chapter will show some ways in which these applications have been made through examples of models of such systems. Of course, one definition of a community is that set of populations of organisms living in a particular place at a particular time. Following this definition, a valid approach to the study of a community is a series of concurrent population studies and a suitable systems model, a set of linked population models. Indeed, in one sense, this is a description of certain "life systems" models, such as those of Gutierrez and his colleagues on cowpea aphids and Gilbert and his co-workers on thimbleberry aphids described in chapter 6. Life systems models "focus" on single species but include key processes involving associated species. An alternative view of communities and, by simple extension, ecosystems, looks not at single species populations and flows of individuals, but at functionally related blocks of organisms which do the same job in the system, and at the associated flows of energy, biomass or nutrients. In simple cases, such as the one described in 8.1, these functional blocks may equate to single species populations, but the fine dynamics of these populations are not the primary concern of the systems ecologist adopting this approach. This **material flow** approach (for want of a better name) has been the central focus of several schools of ecologists based, largely, in North America. The two most prominent of these schools are the so-called **energese** approach of H.T. Odum, mentioned earlier in connection with its symbology for the representation of energy flow in systems, and what I shall call the **IBP** approach, developed for the construction of models of energy or nutrient flow in ecosystems chosen as units of study under the aegis of the International Biological Programme. A detailed example from the work of each of these schools is described in 8.2 and 8.3.

Before presenting the complexities of these studies, however, I shall present an **approach** to a community model based on some of my own work on the fauna of water-filled tree-holes in southern England. I stress now that I have not built such a model nor is the existing data-base sufficient to proceed to such a model at this stage. There is considerable value, however, in showing one way of moving from basic information on the natural history and ecology of a situation to the skeleton of a model, ready for parameter estimation and subsequent validation and experimentation. The community of organisms found in the tree-hole studies is so simple that it provides an admirable subject for an heuristic treatment of this sort.

8.1 Water-Filled Tree-Holes: An Approach to Modelling

8.1.1 Background Information

Tree-holes which contain water are a common feature of forest ecosystems the world over. These habitats are occupied by a variety of insect larvae, crustaceans and micro-organisms, with other groups of animals and plants present in the more complex tropical situation. Much of the fauna is specialized with respect to these habitats and represents larvae of insects found nowhere else. The adults of the species concerned, of course, range much more widely in the woodland or forest system of which the tree-holes form part. I carried out an intensive study of such water-filled containers in beech trees in southern England over a three year period; the detailed results of these studies can be found in Kitching (1971b, 1972a and 1972b). Before the tree-hole environment and the community of organisms it contains can be considered as a subject for a modelling exercise, it is necessary to know a little about the biology of the situation. The following remarks refer to the tree-hole system in the United Kingdom only.

The major source of energy entering the tree-hole system is the vegetable matter falling from the canopy or drifting up from the forest floor. This litter forms a layer of varying depth in the bottom of the hole; it undergoes a gradual breakdown process through the actions of various micro-organisms and through the activities of the macro-fauna which will be considered shortly. On top of this detritus is a layer of free water, resulting from rain running down the trees or entering the holes directly. In periods of heavy rain this inflow may be sufficiently violent to wash out some of the detritus in the tree-holes, and, accordingly, to deplete the basic food source for the organisms that occur in this situation. In the tree-holes I studied, there are commonly found six species of insect larvae which feed directly on this detritus in various forms. They appear to have partitioned the resource, and different species exploit different components of the decaying detritus. Two species may attack the larger particles of detritus or, possibly, the micro-organisms that live on them. These are the larvae of the syrphid fly, *Myiatropa florea,* whose larvae have long posterior breathing tubes which have given them their common name of rat-tailed maggots, and the helodid beetle, *Prionocyphon serricornis,* the larvae of which are crawlers which insinuate themselves among the leaves and twigs of the detritus, grazing on the material they encounter there. When the detritus has undergone a certain amount of breakdown and is available is finer particles, two species of midge are able to use it as food source for their larvae. These are numerically the most abundant of the tree-hole fauna and are the subjects of two of the cited papers (Kitching 1972a, 1972b). The first and more common of these midges is the chironomid, *Metriocnemus martinii* (redesignated *M. cavicola* by later taxonomic work), and the second, the ceratopogonid, *Dasyhelea dufouri.* These two species have larvae of approximately the same size and feeding habit, which suggests that they may compete for the fine-particle detritus in some places at some times. The last two species of insect larvae found in these situations are mosquitoes. Their larvae filter the suspended organic matter out of the water column and hence exploit a separate segment of the vegetable matter in the holes. In my study two species of mosquito occurred but one of them,

Aedes geniculatus, was by far the more numerous. Individuals of the second species, *Anopheles plumbeus,* occasionally turned up, but their occurrence was irregular and infrequent. All of the insect species enter the tree-holes as a result of the actions of ovipositing females, which occur free-flying in the woodland and are attracted to the water's edge in tree-holes in order to lay their eggs. All the species mentioned go through their larval stages in the tree-hole and, except for the syrphid and the helodid, pass their pupal stage in this habitat also. The syrphids and the helodids crawl out of the tree-hole milieu when ready to pupate, and undergo this metamorphosis in crevices in the bark of the tree or dry leaf litter adjacent to the tree-hole. The pupae in the tree-holes eventually emerge and the resulting adults join the population of adults dispersing, feeding, mating and eventually ovipositing in the wider woodland habitat. The rates of development of larvae in the tree-hole situation are dependent upon temperature and hence upon heat input from their surroundings, which in turn will reflect the air temperature around the tree-hole. These are the temperature-dependent developmental processes identified in the earlier discussions of various insect population models. Apart from a few earthworms, probably entering the holes with the incoming leaf litter, the species just listed represent the complete community of macro-organisms occurring in water-filled tree-holes in southern England. There are no predators operating within the system, although in parallel systems in North America this lacuna is filled by the presence of species of mosquito which have predatory larvae (*Toxorhynchites* species). A whole range of predatory insects, mites and even frogs occur in comparable tropical situations. The other major simplifying feature of this and similar situations is that there is no primary production going on within the system. Energy enters the system only in the form of vegetable matter falling from above or drifting from around the tree-holes.

8.1.2 Problem Definition

For the purpose of this exercise, we will imagine that we have the background information on the tree-hole community outlined in the preceding section, together with sufficient quantitative information to proceed to the construction of a model. Such an exercise is justified as an attempt to answer questions such as:

- What is the relative importance of each of the insect species found within the system in terms of energy flow?

- What would be the effect on the community as a whole of the removal of one of the members of the community?
- What would be the effect on the patterns of energy flow through the community following the introduction of a predatory species into the system?
- What would be the patterns of change in the biological community in tree-holes resultant upon particular patterns of rainfall, leaf fall, or adult dynamics?

A great many other questions along the same lines might be approached through the construction of a model of the energy flow through the system. The ones outlined have at some stage been put forward as having relevance in management situations. For instance, in North American tree-holes there exists a pest mosquito, *Aedes triseriatus,* which reaches densities far beyond those recorded for the equivalent *Aedes* species in the situation I studied. It has been suggested that the absence of a species analogous with the midge *Metriocnemus martinii* is the reason that the mosquito achieves these high densities. Accordingly it has been proposed that the introduction of the harmless midge species into the North American system might result in a depression in the numbers of the pest mosquito. A model of the kind proposed might well be used for an evaluation of this management process before it was actually implemented.

8.1.3 Systems Identififation

Following the sequence of processes outlined in chapter 3, the problem must now be tackled of identifying the system of interest, defining the components and processes within it and defining the major flows of material and information. Figure 63 is a Forrester diagram of the tree-hole system based on the background information described above and available in greater detail in the papers to which I have already referred. Eight state variables are identified, each representing the amount of energy tied up in a particular component of the system at any point in time. The detritus and other organic matter in the system are divided into three separate components — large particle detritus, fine particle detritus and suspended organic matter — to correspond with those components which are thought to represent food supplies for particular members of the biological community involved. Of course, this imposes a discreteness upon the continuous process of breakdown of the litter entering the hole, but this is one of the simplifications

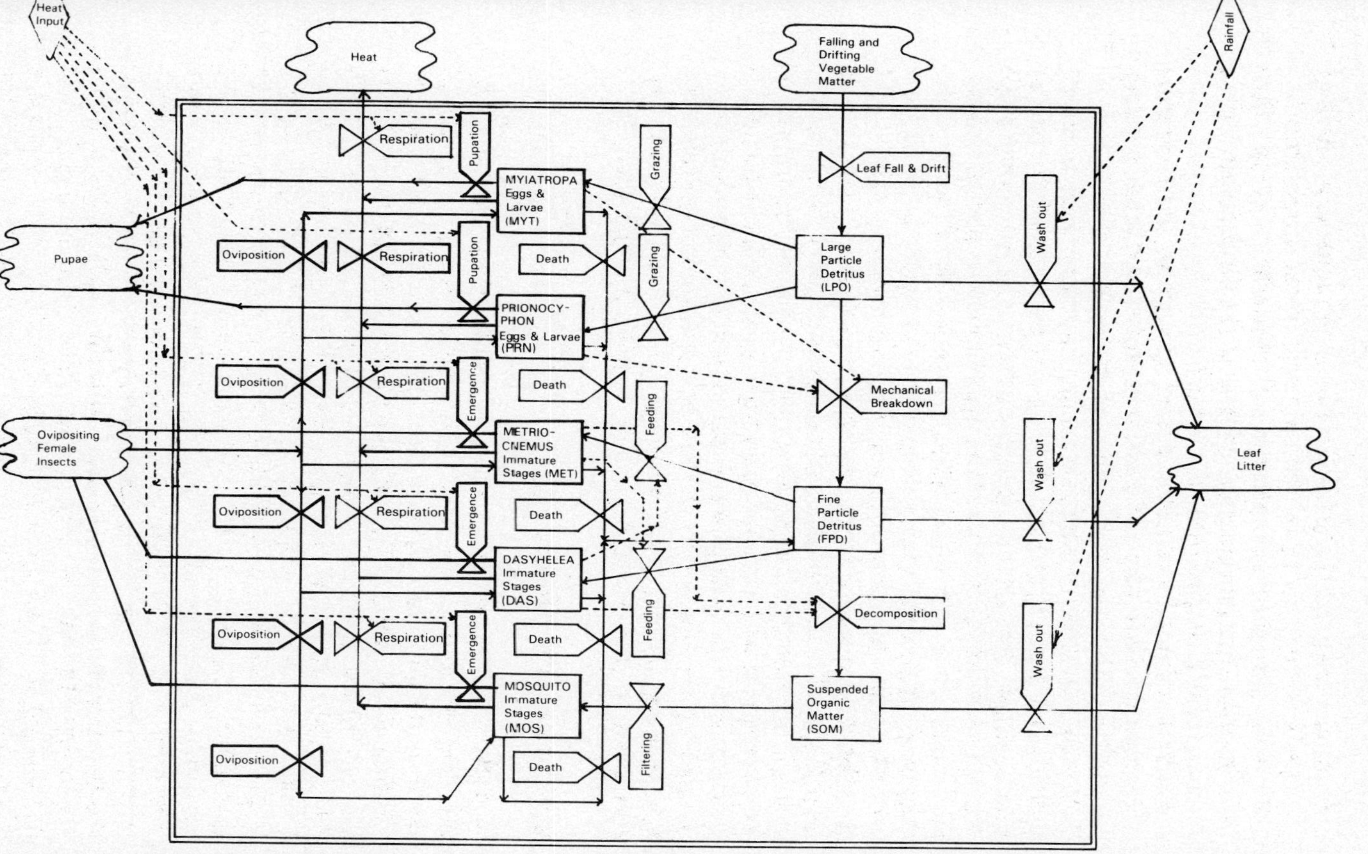

Fig. 63 Forrester diagram of the energy flow through the community within water-filled tree-holes in southern England (compiled from Kitching 1971b)

concomitant with the modelling process in general. The five biological components proposed represent the energy tied up in each of the species of insect in the tree-hole.

All stages of each species occurring in the tree-holes have been lumped together within each state variable to avoid the considerable complication that would accompany any attempt to represent the different age classes of each separately. Also for the sake of simplicity, both species of mosquito are treated as a single state variable. This could easily be made more realistic and divided into two but, bearing in mind the infrequent occurrence of the second species involved, it seems entirely justified to retain the combined quantity. Major energy inputs into the system are represented by the leaf fall and drift processes which cause vegetable matter to enter the hole, and by the process of oviposition, whereby adult females lay eggs at the air/water interface. Energy leaves the system in three basic ways: detritus may be washed out following heavy rain; insects mature and leave the system following their metamorphoses; and energy is lost to the system in the form of heat through the respiration of the organisms contained within the community. The processes of maturation and washing-out already mentioned are direct responses to the action of two other

Table 46. Interaction matrix for state, driving and input variables in the tree-hole model.

From / To	State variables								Driving variables		Inputs	
	LPD	FPD	SOM	MYT	PRN	MET	DAS	MOS	RF	T	LF	EGGS
LPD	–	0	0	–	–	0	0	0	–	–	+	0
FPD	+	–	0	0	0	–	–	0	–	–	0	0
SOM	0	+	–	0	0	0	0	–	–	–	0	0
MYT	+	0	0	+ –	0	0	0	0	0	+ –	0	+
PRN	+	0	0	0	+ –	0	0	0	0	+ –	0	+
MET	0	+	0	0	0	+ –	–	0	0	+ –	0	+
DAS	0	+	0	0	0	–	+ –	0	0	+ –	0	+
MOS	0	0	+	0	0	0	0	+ –	0	+ –	0	+

Key to symbols

LPD – Large particle detritus
FPD – Fine particle detritus
SOM – Suspended organic matter
MYT – Larvae and eggs of *Myiatropa*
PRN – Larvae and eggs of *Prionocyphon*
MET – Immature stages of *Metriocnemus*
DAS – Immature stages of *Dasyhelea*
MOS – Immature stages of mosquitoes
RF – Rainfall
T – Temperature
LF – Leaf fall and drift
EGGS – Insect eggs entering system

variables imposed upon the system from the outside, namely the heat input and the input of rainfall. One other energy pathway is included in the model: that whereby the component representing fine particle detritus is added to, through the mortality process acting upon the populations of each of the insect species occurring in the hole.

Information flows within the system include the effects of the quantities of organisms on the process of mechanical breakdown of detritus, and the possible competitive effects between similar species.

The interactions among the components included in the model and the driving variables which act upon them are summarized as an interaction matrix in table 46. In this matrix, the effect of one variable on another is shown either as positive, negative or zero, according to the nature of the relationship suggested between the pairs of variables involved. As will be seen shortly, it is possible to go straight from this sort of interaction table to a set of equations governing the dynamics of each of the state variables.

8.1.4 Decisions on Model Type

The aim in formulating this particular model is to illustrate clearly and concisely how one approaches the building of a simulation model. Obviously, inherent in this aim is the decision to build a numerical simulation model rather than an analytical, pure mathematical model. In addition, for the sake of furthering the same aim, I shall proceed to a discrete, deterministic model at this stage. As should be clear from treatments of models earlier in this book, it would be possible to go from a model of the simple type being discussed here to one which was couched in continuous terms or to a stochastic version of the same model.

8.1.5 Mathematical Formulation

Once a decision has been made concerning the model type, the nature of the mathematics involved is more or less predetermined. In fact, the discrete nature of the model proposed dictates that the dynamics of the state variables be couched in terms of **difference equations.** In other words, I shall attempt to write equations showing the changes which might be expected to occur in the level of each state variable in a predetermined, finite, time interval.

I have formulated the equations involved in two sets, reflecting basic differences in types of variable. The first of these relates to the dynamics of the **detritus** components and the second, to the **biological** components. The sets of equations involved are set out in full in tables 47 and 48 respectively. These equations are arrived at by a very simple process. Firstly, I consider the changes which will occur in the state of the variable concerned in any time interval in terms of additions and subtractions, I give each of these additions and subtractions a name and a symbol and I write out the basic equation for each variable

Table 47. Basic equations of detritus dynamics in the tree-hole model

Large Particle Detritus

Basic equation

$$LPD_t = LPD_{t-1} - LPDEAT - LPDDEC - LPDWO + LF \qquad (1)$$

where LPDEAT is loss to insect larvae
LPDDEC is loss by decay
LPDWO is loss by washing out
LF is gain by leaf fall – a driving variable read in

Subsidiary equations

$$LPDEAT = f_1\,(MYT, PRN, T) \qquad (2)$$
$$LPDDEC = f_2\,(LPD, T) \qquad (3)$$
$$LPDWO = f_3\,(LPD, RF) \qquad (4)$$

Fine Particle Detritus

Basic equation

$$FPD_t = FPD_{t-1} - FPDEAT - FPDDEC - FPDWO + LPDDEC + INSMOR \qquad (5)$$

where FPDEAT is loss to insect larvae
FPDDEC is loss by decay
FPDWO is loss by washing out
INSMOR is gain through insect larvae dying

Subsidiary equations

$$FPDEAT = f_4\,(MET, DAS, T) \qquad (6)$$
$$FPDDEC = f_5\,(FPD, T) \qquad (7)$$
$$FPDWO = f_6\,(FPD, RF) \qquad (8)$$

INSMOR is calculated in equation 22 shown in table 48
LPDDEC is calculated in equation 3 above

Suspended Organic Matter

Basic equation

$$SOM_t = SOM_{t-1} - SOMEAT - SOMWO + FPDDEC \qquad (9)$$

where SOMEAT is loss to insect larvae
SOMWO is loss by washing out

Subsidiary equations

$$SOMEAT = f_7\,(MOS, T) \qquad (10)$$
$$SOMWO = f_8\,(SOM, RF) \qquad (11)$$

FPDDEC is calculated in equation 7 above.

Note: Symbols as in table 46 except where defined.

Table 48. General and specific equations of insect dynamics in the tree-hole model

A General Equation for the Insect, Species INS

$$INS_t = INS_{t-1} - INSMOR - INSMET - INSRES + INSEAT + INSEGG \quad (12)$$

where INSMOR is loss due to deaths
INSMET is loss due to metamorphosis
INSRES is loss due to respiration
INSEAT is gains from feeding
INSEGG is gain from eggs entering the system

Subsidiary equations

$$INSMOR = f_9 (INS, INSM^a) \quad (13)$$
$$INSMET = f_{10} (INS, T, TIME^b) \quad (14)$$
$$INSRES = f_{11} (INS, T) \quad (15)$$
$$INSEAT = f_{12} (INS, T, \text{Food variable}^c) \quad (16)$$

a. INSM is the species specific mortality rate
b. TIME is the time in day degrees elapsed since the beginning of the season
c. f_{12} will include the state of a second insect variable when interspecific competition for food is involved

A Specific Example for *Metriocnemus* (MET)

General equation

$$MET_t = MET_{t-1} - METMOR - METMET - METRES + METEAT + METEGG \quad (17)$$

Subsidiary equations

$$METMOR = f_{13} (MET, METM) \quad (18)$$
$$METMET = f_{14} (MET, T, TIME) \quad (19)$$
$$METRES = f_{15} (MET, T) \quad (20)$$
$$METEAT = f_{16} (MET, T, FPD, DAS) \quad (21)$$

Summed Mortality for Input to Equation 5 of Table 47

$$INSMOR = MYTMOR + PRNMOR + METMOR + DASMOR + MOSMOR \quad (22)$$

Note: Symbols as in tables 46 and 47 except where defined.

as a simple string of subtractions and additions. Subsidiary equations must then be written for each of the variables which are introduced in the formulation of the basic equation, except where the subcomponent concerned is a driving variable, in which case it will simply be read in from some outside source representing either real data or experimental quantities, the effects of which one wishes to investigate. In writing the subsidiary equations one must list those variables, either state or driving, within the ecosystem which will affect the process represented by the subcomponent that one wishes to define. This subcomponent is the dependent variable on the left-hand side of the equation concerned, and the other components which affect it are the independent variables on the right-hand side of the equations. The form of the function relating the dependent variables to the independent variables is not defined in the equations I have written but is represented simply in general form by a series of f_i notations.

The set of five equations representing the dynamics of the insect components are all very similar in form, hence I have written a single general insect equation in the first part of table 48, identifying the basic changes in the energy tied up in a particular component. The equations for particular species of insect can then be written by slight modification in the notation from this general form, and I give a specific example in the table.

Two additional features of the set of equations represented in the two tables are noteworthy.

1. Some independent variables entering into the basic equations of change of state are in fact calculated elsewhere in the set of equations that I have just formulated. This means that the model needs careful structuring so that particular values are available to use in calculations at the right point in the flow of computation, which will be defined in the model.

2. In the insect equations there is a single set of mortality rates which are parameters, that is constants, which we can imagine as being characteristic numbers for each organism concerned. Before the model can be used, these numbers will either have to be measured experimentally or be themselves the subject of experimentation using the model.

The next stage in the process of mathematical formulation would be to put some characteristic form on each of the functions concerned. In some instances as with the mortality and decay functions, obvious possibilities come to mind — notably various negative exponential forms such as the one explained in detail in association with the flour beetle model described in chapter 6. In other cases, as with equation 10, which represents the filter-feeding of mosquito larvae, the relationship may be a very simple one involving a constant rate of consumption, such that the amount of suspended organic matter removed by the mosquito larvae may be wholly dependent upon the elapsed time, in a simple linear fashion. In some cases, however, there is no simple form that is immediately apparent, and two courses of action are open to fill the needs of the model.

1. Statistical analyses can be carried out on the data base, in order to obtain predictive equations relating the level of the dependent variable to that of the independent variables. This would be particularly important in those instances, such as equations 6 and 21, where there are several independent variables in the equation. This is not the place to go into the

theory and background of regression analysis; suffice it to say that whole books have been written on the subject (for example, Draper and Smith 1966, Li 1964). However, it is necessary to point out that an equation obtained by these means, although useful and functional in terms of the modelling context, has no theoretical base within the biology of the processes involved. It is purely a statistical description of the pattern present in the data base, and should perhaps be substituted, wherever possible, by a form of equation that has a biological basis. However, in the modelling situation this is almost always impossible, at least for some of the relationships required in order to complete the model, hence, the techniques of regression analysis remain a vital part of the systems ecologist's armoury.

2. The second technique of filling out the equations is considerably more complicated and will be mentioned only in passing. This is where the equation as written is in fact a summary of a great deal more information, and depends for its implementation on the inclusion of yet more subsidiary equations. The best example from the current case is equation 14 in table 48, which represents that set of insect equations describing the process of metamorphosis whereby organisms leave the tree-hole system as a result of maturing to a non-aquatic stage. The independent variables identified here are the actual abundance of the insects, the ambient temperature, and the elapsed time from the beginning of the particular episode of population development. This means that the insects involved have accumulated the appropriate number of day degrees and food input in their history. This reflects the process of maturation described in 6.1 and implies a very complex set of processes dependent upon the age of the organisms concerned. In the present context this would be a very strong candidate for the development of a submodel to form part of the larger ecosystem model, perhaps reflecting the output from a series of population models.

8.1.6 Decisions on Computing Methods and Programming

The couching of the dynamic equations of the tree-hole system in difference form makes the transition from mathematical formulation to computer programme a relatively simple one. The model could be programmed in a number of languages suitable for the representation of discrete changes, including FORTRAN, ALGOL or one of the discrete simulation languages such as

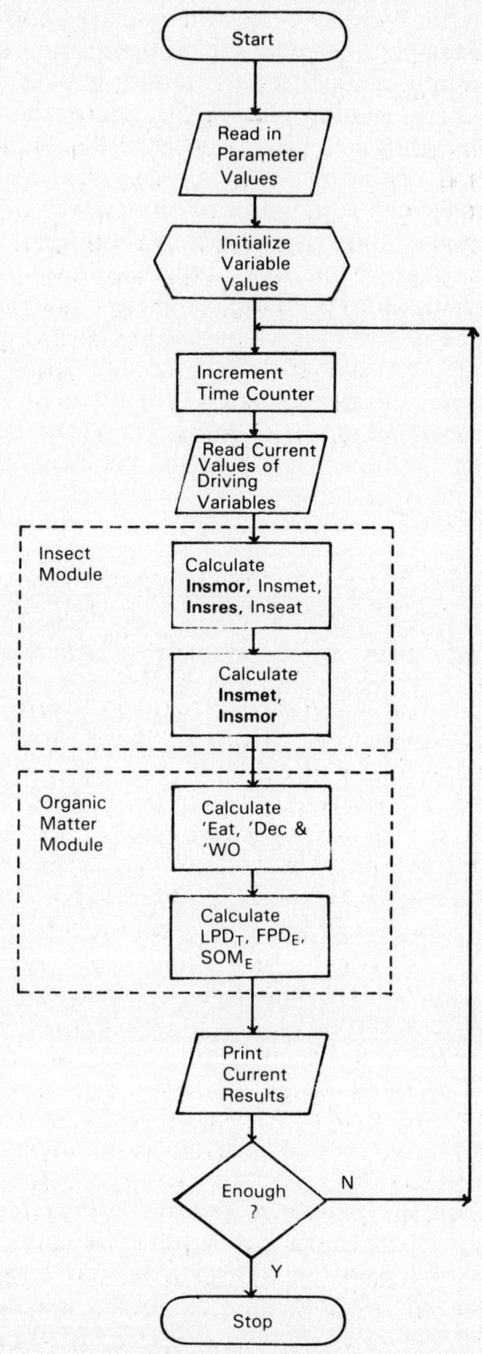

Fig. 64 Flowchart of a tree-hole community model. See text and tables 46 and 47 for further explanation and key to abbreviations

SIMULA or GASP. The major pitfall in the structuring of the computer programme would be the sequence of calculations made within the major time loop, which would be the central feature, as in most ecological models. Figure 64 shows a flowchart which represents a first approach to the programming of this particular model. It will be seen that the flow of computations identified lends itself to structuring at the actual programming phase with two major subroutines, the insect module and the organic matter module, corresponding to the two basic types of variable. Once a decision had been made concerning the form of the functional relationships referred to generally in tables 46 and 47, it would be a straightforward matter to proceed from the scheme laid out in figure 64 to the formulation of a computer programme.

8.1.7 Further Steps

At this stage I shall leave the tree-hole model, having shown in some detail how to approach the construction of a community model of this type. The form of equations that have been identified indicates those data which would be needed to proceed further. Even though a detailed study of the system has been carried out and is described in the papers referred to earlier many of the quantities identified in the equations do not figure in the data base, primarily because it was not collected with modelling in mind. Having proceeded this far, we have in fact identified what we do not know, and a whole sequence of laboratory and field work can now be designed in order to elaborate upon the functional relationships involved and to obtain the parameter values concerned in these relationships. These processes of function and parameter estimation would be the next step in our modelling efforts; they would be followed by a validation procedure which, most obviously, would comprise comparisons between the sequence of state vectors, which could be generated by the model given a sequence of real data as driving variables, and the actual state vectors measured over a period of time in the field situation. Undoubtedly, the model would have shortcomings initially and the cyclical procedure of model improvement described in chapter 3 would have to be implemented. Eventually we could arrive at a satisfactory model, and enter into the processes of experimentation by which we would attempt to answer the questions formulated initially in our justification for the whole modelling effort.

8.2 The Grassland Biome: An IBP Compartment Model

8.2.1 Background Information

The International Biological Programme was a period of co-ordinated research activity in the biological sciences which ran from about 1965 through to 1974. Ecological activities were a central feature of this programme and focused on the comparative study of biomes in different continents and in different parts of the same continent. In the United States, this activity was under the auspices of the Analysis of Ecosystems Program of the National Science Foundation and directed a massive amount of attention, both in terms of data gathering and analysis, at a series of ecosystem types within North America. These **biomes** included tundra, coniferous forest, eastern deciduous forest, desert and grassland. The use of complex systems modelling techniques was adopted by this programme for the synthesis stage of the study involved. The volume of papers by Patten (1975) summarizes much of this work. The models built have been of a distinct type which, although not wholly restricted to the IBP, have reached their fullest development under its umbrella. It is therefore entirely appropriate that one of the examples of ecosystem models examined in detail in this chapter should be drawn from the suite of models that has been generated by this programme.

Before tackling a particular example of the genre, a couple of qualifications are necessary so that this work may be placed in the context of other models of various levels of ecological integration, some of which have been described already.

1. The models built as part of this coordinated effort are very complex. They are wholly synthetic and build submodel upon submodel, each of which contains from a few to many state variables. This means they are difficult to interpret and even more difficult to validate. They have been used as a way of putting together a very large and varied amount of information existing on a particular ecosystem.

2. Few of them have been published to date, which in practical terms is more restrictive than their complexity. The output from the various research groups has taken the form of working papers, internal reports and unpublished symposium proceedings. The reports and working papers which are available generally carry the preamble that they are not to be quoted as published work and are to be regarded as interim documents only. Patten's work (1975) contained summaries

only of models, and few of them can be reconstructed or evaluated in any detail from the information presented in that volume by the authors concerned. This situation is being rectified as the series of synthesis volumes appears. These include, to date, works on arid ecosystems (Goodall and Perry 1979), grasslands (Coupland 1979, Breymeyer and van Dyne 1980), forest ecosystems (Reichle 1980) and tundra ecosystems (Bliss, Heal and Moore 1981).

The first model to be built and also the first to be fully published was that simulating the **grassland** biome. This project drew upon data from eleven sites in North America, where extensive research was carried out into the variety of grassland types represented on that continent. A mass of information on nutrient and energy flows in the inorganic and organic segments of the various grassland ecosystems involved was accumulated; it was from this data that a team of workers produced a series of simulation models. The most extensive of these models, which will be described here, goes under the name, **ELM**. The processes of work involved are described in a series of publications, including those of Innis (1975), van Dyne and Anway (1976) and a major synthesis volume under the editorship of G.S. Innis published in 1978. This particular model can now be regarded as adequately documented and can be evaluated accordingly. The extensive nature and high degree of complexity of the model make it difficult to present anything approaching a complete picture of its structure and functioning in summary form, so I shall restrict myself to an overview of the whole model, with a detailed account of two of the submodels, those associated with primary production. These two submodels, the **plant phenology** and the **carbon-flow** models, simulate the dynamics of the key plant groups in the grassland. The same headings will be used as in 8.1 to reflect the various stages in model development.

8.2.2 Problem Definition

The problem definition phase of this work has already been foreshadowed in introductory remarks. The ecologists involved in the grassland modelling attempt were trying to put together a description of a general grassland system that could be used to make valid and extensive comparisons between sites, as well as investigating the dynamics of the system at any particular site. They wished to investigate the effects of weather factors upon the grassland system, its producers and consumers. In addition, they particularly wanted to be able to model different sorts of flows within the system, not just of energy but also of various key nutrients such as nitrogen and phosphorus, together with water and, to some extent, biological components. Such a model, completed, would have obvious value in investigating management options, as well as in the identification of key processes or organisms.

Innis (1978) points out the open-ended nature of the problems which can be tackled with a model of the sort built, but explains how the actual questions addressed were narrowed down to four areas:

1. What are the effects on primary production of variations in levels of grazing, temperature and water regime, and through the addition of nitrogen or phosphorus?
2. What is the effect of these changes on the carrying capacity of the ecosystem?
3. Are model results consistent with field-data (that is, can the model be validated)?
4. What are the effects of the various pertubations listed under (1) upon the producers?

8.2.3 Systems Identification

Innis (1978) presents a modified Forrester diagram of the complete grassland model, and I have redrawn this very complex figure as two separate diagrams in figures 65 and 66. Figure 65 shows the basic biological submodels of the grassland system, comprising a **producer** submodel, an **insect consumer** submodel, a **mammalian consumer** submodel and a **decomposer** submodel. The flows involved are in terms of biomass (or grams of carbon), representing the basic energetic transitions within the system. This diagram attempts to show one other essential and characteristic feature of this model, the "three-dimensional" nature of the state variables involved. Many of the state variables represented by the boxes in the diagram, instead of being single quantities, are in fact sets of quantities representing different species, different age groups, different layers in the soil and so forth. These additional dimensions are variable in number and are one reason for the great complexity of the final model. The flows shown in the diagram therefore are in a sense a two-dimensional slice of the complete model which has other slices representing the other layers just mentioned.

These submodels are all richly connected within themselves, but also have connections with other submodels illustrated in

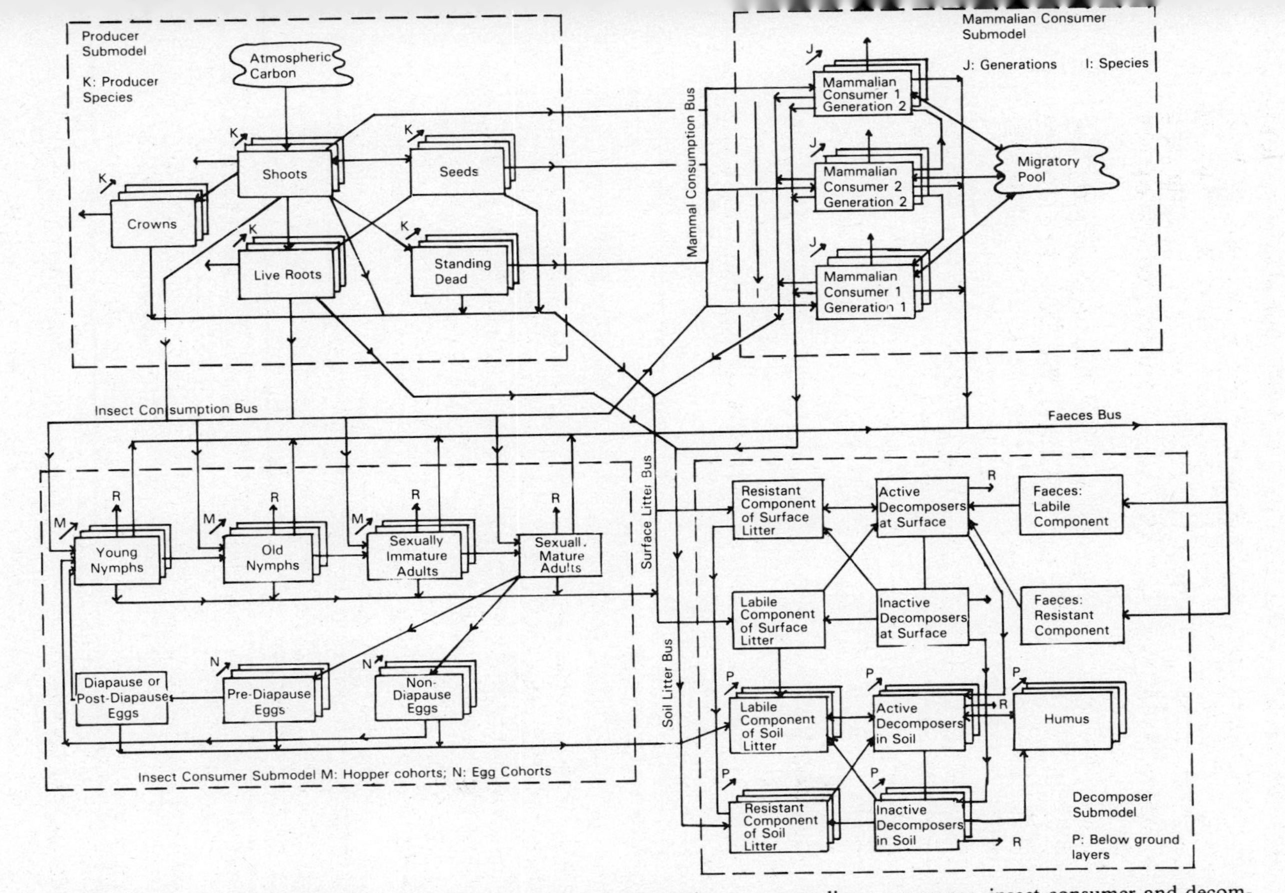

Fig. 65 Flowchart of the ELM grassland simulation model: the producer, mammalian consummer, insect consumer and decomposer submodels (modified from Innis 1978). R – represents respiratory loss

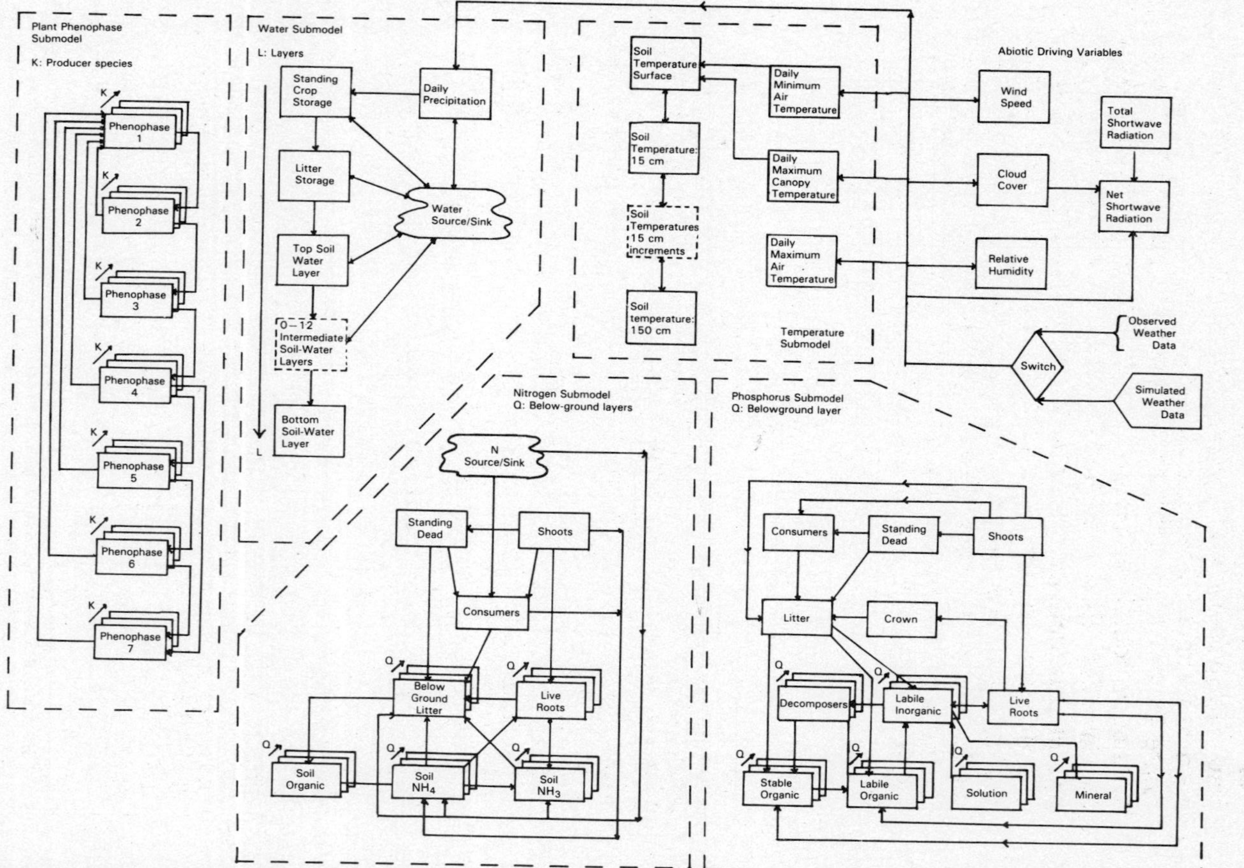

Fig. 66 Flowchart of the ELM grassland simulation model: the plant phenophase, water, temperature, nitrogen and phosphorus submodels (modified from Innis 1978)

figure 66. These are of a more subtle nature, and include input from the **phenology submodel** governing the age structure of the plants involved in the producer submodel, input of the levels of nutrient feeding to a variety of biological components shown in figure 65 and input from the **temperature submodel** to those submodels containing temperature-dependent processes. Figure 66 includes three submodels, the **water, nitrogen** and **phosphorus** modules, which allow an investigator to look at the overall grassland systems from a point of view different from that of the **carbon flow** dynamics on which the biological submodels are based. Thus, for example, an investigator interested in nitrogen dynamics is able to use the nitrogen submodel for experimenting with the system variables. Overall the variety of

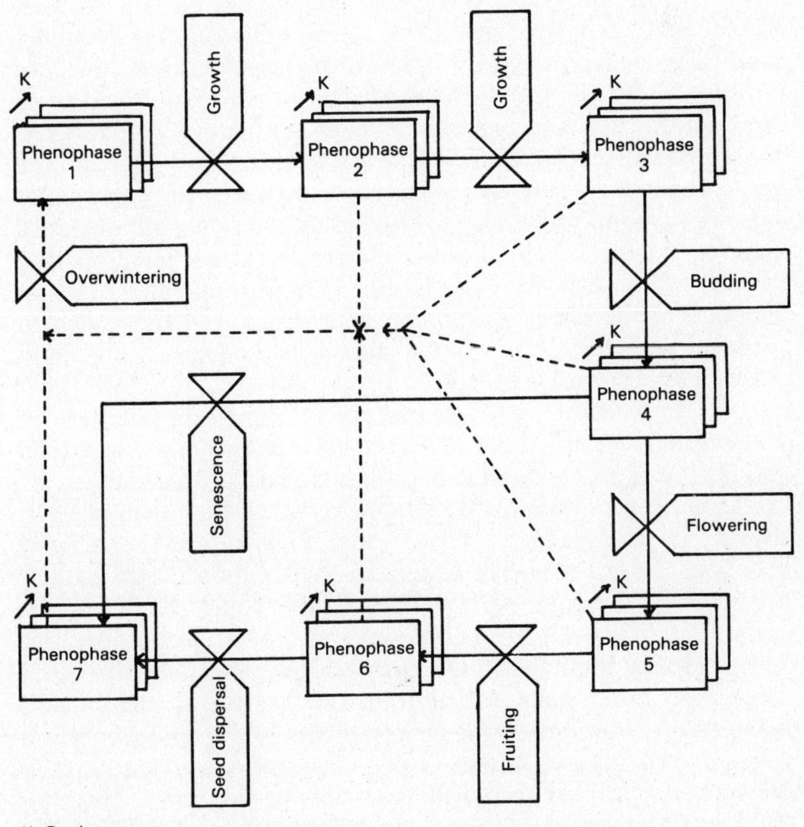

Fig. 67 Flowchart of the plant phenology submodel of the ELM grassland model (after Sauer 1978)

submodels available reflect the multifaceted nature of this approach to the study of systems dynamics. Figure 66 also shows the basic driving variables involved in the ELM model, and it will be observed from the figure that these can take the form of the observed weather data for a particular period of time or simulated patterns of weather, and the choice between these two paths will again depend upon the nature of the questions being asked at any particular time.

Turning from the overview of the model represented by figures 65 and 66, I will now examine the structure and systems identification phase of the two submodels singled out for special attention in this account.

1. The **plant phenology submodel** is shown in a fuller form in figure 67. Here, the model builders divided the vegetation of particular sites of interest into a number of major classes, and simulated the processes of ageing and maturation for each of these as shown in the figure. For the Pawnee site, for example, the five groups involved were the warm season grasses, the cool season grasses, the forbs, the shrubs and the cactus. In order to represent the dynamics of these groups, the life cycle of each was considered in seven stages. These are listed as part of figure 67 and range from winter quiescence through the period of germination and growth up to flowering, fruiting, seed dispersal and senescence. Each of these classes is connected by a rate process as shown in the figure. One further phenomenon included in the model is indicated by the dotted flows, which represent the process by which plants at any one of a number of stages can become quiescent.

2. The **carbon flow model** of primary production is illustrated in more detail in figure 68. Once again the model divides the primary producers on any one site into a number of categories, and simulates the carbon dynamics for each class as shown in the figure. Three components and associated processes, which properly belong in the decomposition model, are also included in the figure to show some of the major connections between this submodel and that of decomposition. The processes involved in the carbon flow model are the usual ecological ones such as reproduction, growth, death and respiration. Similarly, the components involved are those usually recognized in studies of primary producers, representing the standing crop of seeds, shoots, crowns, live roots, standing dead and atmospheric carbon surrounding the system.

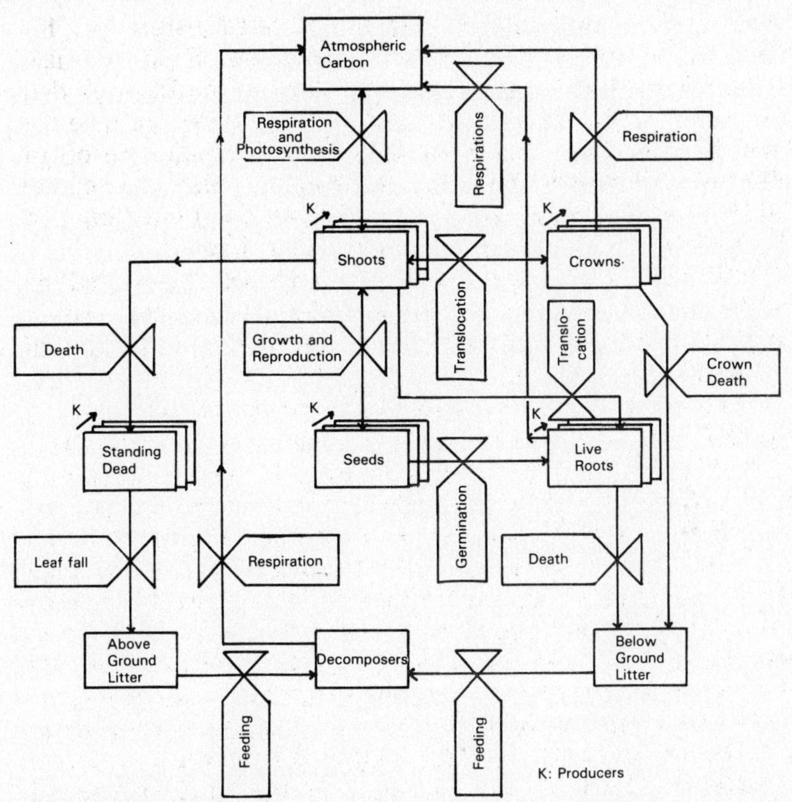

Fig. 68 Flowchart of the primary production submodel of the ELM grassland model (after Sauer 1978)

8.2.4 Decisions on Model Type

Some of the decisions involving the type of model to be built in the grassland simulation are implicit in the comments already made. Obviously, in a system of this complexity, any hope of succinct mathematical representation in a form amenable to further analysis must be set aside. A large number of transition equations will be involved, based to a large extent on empirical relationships derived from field observations or the literature. Once the decision has been made early in the programme to build a complex simulation model, further decisions are needed about its nature. It was decided to build a deterministic rather than a stochastic model; in the interests of simplicity, this seemed the obvious choice. It was also decided that the model should be built to represent the dynamics of the system in a discrete manner using time steps of one to seven days.

8.2.5 Mathematical Formulation

The total ELM model contains over one thousand state variables, when the layered nature of the model as shown in figures 65 and 66 is taken into account. The transitions among these state variables are many and complex, and cannot all be described in a summarizing account such as this. However, the flavour of the model can be conveyed by close examination of the transition equations involved in the two submodels, the **plant phenology model** and the **carbon flow model** for primary production. Even with this restriction, there are more than a

Table 49. Transition equations in the plant phenology submodel of ELM (after Sauer 1978)

Vegetative Transitions [f(1,2), f(2,3), f(3,4)]

$$f(x,y) = r_1(x)\ P(x)\ \alpha(t_1)\ \beta(e_w)\ \delta_t$$

where

$$\alpha(t_1) = h[p_x, p_y, 0, 1, t_1]$$
$$\beta(e_w) = 0.5[\sin(\pi\ e_w) + 1]$$

Symbols:

$r_1(x)$ – phenophase transfer coefficient
$P\ (x)$ – phenophase contents
$h(x_1, x_2, y_1, y_2, x]$ – an s-shaped piecewise linear function
e_w – effect of soil-water potential and root distribution (see below)
t_1 – 10-day moving average of product of the average photoperiod air temperature and insolation
p_x, r_y – values of t_1 for phenology change
δt – time step in days

Other Transitions

$$f(4,5) = r_1(4)\ P(4)\ \alpha(t_1)\ e_w\ \delta_t$$
$$f(5,6) = r_1(5)\ P(5)\ \alpha(t_1)\ \delta_t \qquad (e_w \leq 0.8)$$
$$f(6,5) = -r_1(5)\ P(6)\ \delta_t \qquad (e_w > 0.8)$$
$$f(4,7) = r_1(4)[P(4) - f(4,5)]\alpha(t_1)(1 - e_w)\ \delta_t$$
$$f(6,7) = r_1(6)\ P(6)\ \alpha(t_1)(1 - e_w)\ \delta_t$$

Calculation of e_w

$$e_w = \sum_{j=1}^{n} a\ w_1(j)$$

where

$$a = \begin{cases} 1 - \psi_j/w_2 & , \psi \leq w_2 \\ 0 & , \psi > w_2 \end{cases}$$

Symbols:

ψ_j – soil-water potential in each of n soil strata
w_1 – the normalized distribution of water absorbing capacity
w_2 – the permanent-wilting soil-water potential.

---

dozen major equations of change and many more subsidiary equations to back them up.

The equations involved in the **plant phenology submodel** are given in table 49, derived from the account of Sauer (1978); they show that there is one basic equation for the first three transitions in the model illustrated in figure 67. These are transitions between vegetative stages of growth, and the equation relates the rate of change, through a transfer coefficient, to the contents of the particular phenophases involved, to the day-time air temperature averaged over the previous ten days, and to a variable which summarizes the effects of the soil-water potential and associated root distribution. The other transitions involved in this submodel are between the reproductive and senescing stages of the particular segment of vegetation being considered at any time, and the five equations involved are shown separately in table 49. Again, they have a form comprising a transfer coefficient acting upon the current phenophase contents modified by recent temperature conditions. Beyond this, however, they differ in minor ways from each other, including as they do the effect of soil-water potential and root distribution in a variety of different forms. The table concludes with the equation used for the calculation of this soil-water potential and root distribution effect. It is based upon the soil-water potential measured in each of a number of soil layers and the distribution of water-absorbing capacities of the roots for the different producers involved, together with the value of the soil-water potential which leads to permanent wilting. One exchange shown in the model but not given in table 49 is that which governs the end of senescence. This is achieved in the model by transferring the contents of all phenophase variables into the first phenophase variable. This simulates the state of the system at the beginning of the growing season, for example. As in previous models discussed, such a transfer of material must be compensated for by setting the other phenophase variables to zero after the event.

A much more complex situation is encountered in the formulation of the **carbon flow model**, which is set out, in part at least, in tables 50 and 51. Table 50 gives the expressions used by the ELM modellers for each of the transitions shown in figure 68. These take a variety of forms and most of them speak for themselves. The mathematics involved are essentially simple and the equations are all couched in linear terms, the essential nonlinearity of many of the functions being circumvented mathematically by the use of piece-wise functions. This means that, instead of using a non-linear equation to represent, say, an

Table 50. Transition equations in the carbon flow primary production submodel of ELM (after Sauer 1978)

Root Death

$$f = \begin{cases} r_9\, R\, a_2 & \text{– Perennials} \\ r_9\, R\, a_2\, a_3 & \text{– Annuals} \end{cases}$$

where $a_1 = 1 - e_w$
$a_2 = h[0.75, 1.5, 0.9, 1, a_1]$
$a_3 = 1 - e_a$

Symbols: as in table 49 except for
e_a – effect of phenology
R – current weight of live roots
r_9 – the root death coefficient

Shoot Death

$$f = \begin{cases} Tr_7(a_2 + a_3) & t_5 > t_{13} \\ T/4 & t_5 \leq t_{13} \end{cases}$$

where $a_1 = 1 - e_w$
$a_2 = h[0, 0.8, 0.8, 1, a_1]$
$a_3 = h[5, 7, 0, 0.2, M]$

Symbols: as before except for
T – current weight of live shoots
M – the mean phenophase
r_7 – the shoot death coefficient
t_5 – the daily minimum temperature
t_{13} – the maximum temperature for frost damage to live shoots

Crown Death

$$f = \begin{cases} r_6\, B & \text{– Perennials} \\ r_6\, B(1 - e_a) & \text{– Annuals} \end{cases}$$

Symbols: as before except for
B – the current weight of live crown
r_6 – the crown death coefficient

Fall of Standing Dead

$$f = r_{10}\, D(a_1 + a_3)$$

where $a_1 = h[c_{11}, c_{12}, 0, c_{10}, w_4]$
$a_2 = 20{,}000\, w_5/(T_t + D_t)$

$$a_3 = \begin{cases} a_2 & a_2 \leq c_{10} \\ 1 - c_{10} & a_2 > 1 - c_{10} \end{cases}$$

Symbols: as before except for
D – the current weight of standing dead
r_{10} – the standing dead/litter coefficient
c_{10} – relative effect of precipitation on water content
c_{11} – smallest storm size causing fall
c_{12} – largest storm size causing fall
w_4 – daily precipitation
w_5 – water content of the plant canopy

Table 50. (cont.) Transition equations in the carbon flow primary production sub-model of ELM (after Sauer 1978)

Seed Production

$$f = r_2 T \left[\frac{P(5) + P(6)}{T} \right]$$

Symbols: as before except for
r_2 – the shoot to seed transfer coefficient

Seed Germination (apportioned equally to roots and shoots)

$$f = \begin{cases} 0 & a_2 < -5 \text{ and } t_{10} < t_9 < t_{10} + 10 \\ a_3 & a_2 \geq -5 \text{ and } t_{10} \geq t_9 \geq t_{10} + 10 \end{cases}$$

where
$$a_1 = (t_9 - t_{10} - 2.5)/10$$
$$a_2 = (\psi_1 - \psi_2)/2$$
$$a_3 = S\, r_3(1 + \sin 2\pi\alpha)/2$$

Symbols: as before except for
r_3 – the seed to shoot/root transfer coefficient
S – the current weight of seeds
t_9 – the soil-surface temperature
t_{10} – the minimum temperature for seed germination

Translocation to Roots and Crowns

$$f = \text{Max}\,[a_1, a_2, a_3, a_4, a_5, a_6]$$

where
$$a_1 = \begin{cases} c_5 & e_p \geq .1 \\ c_6 & e_p < .1 \end{cases}$$
$$a_2 = \begin{cases} c_5 & e_N \geq .1 \\ c_6 & eN < .1 \end{cases}$$
$$a_3 = c_5 + (1 - c_5)L/n$$
$$a_4 = \begin{cases} h[4, 7, c_5, c_6, M] & \text{– Perennials} \\ c_5 & \text{– Annuals} \end{cases}$$
$$a_7 = c_7\, T/R$$
$$a_5 = \begin{cases} a_6 & a_7 < c_6 \\ a_7 & a_7 \geq c_6 \end{cases}$$
$$a_6 = \begin{cases} c_5 & e_w \geq c_6 \\ c_6 & e_2 < .2 \end{cases}$$

Symbols: as before except for
c_5 – the minimum fraction of photosynthate leaving shoots
c_6 – the maximum fraction of photosynthate leaving shoots
c_7 – the root/shoot ratio at equilibrium
e_N – the effect of nitrogen
e_p – the effect of phosphorus
L – live shoots removed by grazing
n – the net photosynthetic rate

Table 50. (cont.) Transition equations in the carbon flow primary production submodel of ELM (after Sauer 1978)

Crown Respiration

$$f = \begin{cases} 0 & t_9 < t_{12} \\ a_1 B r_5 & t_g \geq t_{12} \end{cases}$$

where $a_1 = [(t_9 + t_{11})/2 - t_{12}]$

Symbols: as before except for

- t_{11} – the soil temperature at 15 cm down
- t_{12} – the minimum temperature for crown/shoot flow
- r_5 – the crown respiration coefficient

Shoot Respiration

$$f_{SR} = \begin{cases} r_{11}\, g\, e_g & t_6 \leq t_3 < t_7 \\ r_{12}\, g\, e_g & t_7 \leq t_3 < t_8 \end{cases}$$

where $e_g = 1 + 0.25\, L/(c_4\, n)$

Symbols: as before except for

- r_{11}, r_{12} – different shoot respiration coefficients
- g – the gross photosynthesis rate (see table 51)
- n – the net photosynthesis rate (see table 51)
- c_4 – rate or grazing removal and net photosynthesis which will cause an untouched pasture to become overgrazed in three years
- t_3 – the average photoperiod canopy temperature
- t_6 – the minimum temperature for photosynthesis
- t_7 – the optimum temperature for photosynthesis
- t_8 – the maximum temperature for photosynthesis

Table 51. Photosynthetic rate equation and related expressions in the carbon flow primary production submodel of ELM (after Sauer 1978)

Gross Photosynthesis

$$g = 1.032 \min\,[e_w, e_t, e_N, e_p\, e_a] \min(i_1, i) i_2\, Q$$

Symbols: as in tables 49 and 50 except for

- e_t – the effect of temperature
- i_1 – the light saturation for photosynthesis
- i – the incident light
- i_2 – the fraction of intercepted light available
- Q – the quantum efficiency

Net Photosynthesis

$$n = g - f_{SR}$$

Symbols: as before.

The Effect of Temperature

$$e_t = h[t_3]$$

where

$$t_3 = t_4(1 - \cos a)/a + t_5$$

$$a = l\pi(l + 4)$$

Symbols: as before except for

- t_4 – daily minimum temperature
- l – photoperiod length

S-shaped curve, the function is divided into a series of segments, one following on the other, each segment being linear, but the combination of segments approximating to the non-linear shape apparent in the data. This leads to computational simplicity but is heavier on programming. In addition to the transfer equations representing the rate processes illustrated in figure 68, the carbon flow model also computes gross and net photosynthesis. The equations by which this is done are shown in table 51, they are of similar form to the equations discussed previously and, indeed, use intermediate variables computed elsewhere in the model for other purposes. All of the equations in the carbon flow model are based on data available to the modellers and on functional forms, derived either from simple comparison of the data with standard formulae or using models that have appeared in the botanical literature for part or all of the processes involved. References to these sources and a more extended account of the derivation and functioning of the equations concerned can be found in Sauer (1978).

8.2.6 Decisions on Computing Methods and Programming

As Innis (1978) explains, the actual modelling process was carried out in a way which was somewhat unusual but admirably suited to the great size of the ELM model. Essentially, each of the submodels identified in figures 65 and 66 was the responsibility of a separate subgroup of workers and was later synthesized into the overall simulation model. Rather than use one of the standard languages for the modelling of the grassland ecosystem, the modellers chose to write their own simulation language, which they called SIMCOMP. It was developed by Gustafson and Innis and is described by the former in the 1978 volume. SIMCOMP is a discrete, FORTRAN-based simulation language; like most simulation languages it enables the simulation modeller to short cut many of the programming problems associated with a standard higher level language. In particular, SIMCOMP allows a single parameter declaration for all parts of a structured programme, a large capacity for state variables and flows among state variables, the ready inclusion of user-defined functions and subroutines, the automatic display of tables and graphs for state variables and their dynamics, the display of tables of flow rates and, lastly, a free format for the input of data. The compiler for the simulation language was prepared by members of the ELM team and was used throughout the modelling process. The compiler they prepared and the simulation language that it implements have received wide acceptance and have been used now in a number of other projects in a variety of institutions.

8.2.7 Parameter Estimation and Sensitivity Analysis

In the version of the ELM model which is written up in the 1978 volume edited by Innis, the parameter values used are those appropriate to the Pawnee site, chosen out of the range of sites involved in the IBP grassland ecosystem project. To quote the chapter by Steinhorst *et al.* (1978): "The many parameters of ELM are known with varying degrees of precision. Some have long been established in the literature; others are estimates by grassland researchers; still others have been measured empirically in grassland biome experiments". Some of these values are listed by Steinhorst *et al.* as introduction to their description of sensitivity analysis carried out on the model. This sensitivity analysis was designed to identify which of the many parameters were critical in determining the nature of the output from the model overall, but served the secondary purpose of permitting objective comparison of two basic methods of sensitivity analysis:

1. The first method was described by Tomovic (1963), and is based on the calculation of **partial derivatives** from differential equation models of systems. In simple terms, this means calculating how much one can attribute change in the output of the model to change in particular parameters within that model.
2. The second approach to sensitivity analysis was described by Steinhorst and his colleagues as the **perturbation of parameter groups**, and the steps involved in this analysis are laid out in table 52. Unlike the Tomovic approach this

Table 52. Sensitivity analysis and the ELM model – the perturbation of parameter groups (after Steinhorst et al. 1978)

Steps Involved
1. Identification of parameters to be analyzed and output variables to be used in assessment
2. The grouping of these parameters into macro-parameters
3. The subsampling of combinations of these macro-parameters using a fractional factorial design (see Shannon 1975)
4. Running the model with unperturbed and perturbed parameter values in the various combinations selected under (3)
5. Comparison of results and assessment of the significance of the response using the F-statistic

one is based on pragmatic considerations supported by experimental design. Essentially, parameters whose effects and roles in the model are to be analyzed must be identified and, under certain circumstances, such as when there are a great many of these parameters, they must be grouped into what the authors call **macro-parameters**. Where the combinations of these macro-parameters are too great to be evaluated in total, a subsampling of them is made using a fractional, factorial design (see Shannon 1975 for details). The model is then run with both unperturbed and perturbed values for the macro-parameters and the various combinations of these and the results are compared using the F-statistic.

For the ELM model, the choice of parameters for sensitivity analysis was made, at the request of Steinhorst and his colleagues, by each of the workers associated with particular submodels. They were asked to identify a short list of variables from their areas of interest, dividing this list into those variables which they thought would have a significant effect on the output of the model; the remaining parameters were to be chosen at random. The reasoning behind this decision was that, if those variables considered likely by the research workers, should indeed turn out to be of particular importance, then this would be in a sense a validation of the model structure. Steinhorst and his colleagues considered in turn both the approaches to sensitivity analysis described above. They concluded that there were a number of computational and mathematical problems associated with Tomovic's approach which made it quite impractical for application to a model of the type and size being considered. Although they express the opinion that the method was useful for the close examination of small systems or perhaps subsystems of more complex structures, it was inapplicable to the ELM model. The second method, involving the pertubation of macro-parameters, was the one they chose to use and their 1978 paper presents the results of this analysis in detail. They conclude:

> Of the parameters to which the model is sensitive, the abiotic ones seem to be frequently important but not uniformly. Actual evapotranspiration was statistically very sensitive, but no meaningful biological fluctuation occurred. Both primary production and e_C [carbon-dioxide production from the soil] responded to parameter changes. The fractional factorial ANOVA leads us to a number of conclusions about production, but not about e_C. Since soil respiration is affected by abiotic, producer and consumer changes, the effects of macro-parameters are less separable.

They also found that those parameters, which had been identified by the research workers as likely to be significant in determining the output of the model, did indeed produce more major changes when perturbed in comparison with that group of parameters chosen at random. This validation leads us to the next stage in the modelling process, that of the validation process itself.

8.2.8 Validation

In addition to the validation which can be deduced from the sensitivity analysis described in the previous section, each submodel was tested by the modellers against appropriate field data. The correspondence between model output and field data varied from model to model, from site to site and from season to season. Figures 69 and 70 present some of these results from the primary production submodels. Those presented in figure 69 compare simulated and observed data for warm season grasses on the Pawnee site for the period 1970 and 1971(a), and a similar comparison is made for 1972(b). Figure 70, in contrast, shows comparisons between the output of the model with respect to the standing crop of the five major groups of primary producers over time under six different sets of experimental conditions. These experiments were run in the computer and on the ground and the two sets of results were compared. It will be seen from these two figures that where the dead and live shoots are combined, there is a fair agreement between observation and prediction. However when live and dead shoots are separated, as in the second set of curves in figure 69, then there is far less satisfactory agreement. Sauer (1978) and Woodmansee (1978) discussed these results and the reasons for the deviations between the predictions of the model and field observations in considerable detail.

8.2.9 Experimentation

The use of the model for testing management options and answering questions about the functioning of the grassland ecosystem was foreshadowed in the previous section. In the introductory paper describing the model, Innis (1978) lists a longer sequence of thirteen model experiments which were run in the computer; the results from six of these appear in figure 70. The treatments involved the addition to the system of water or

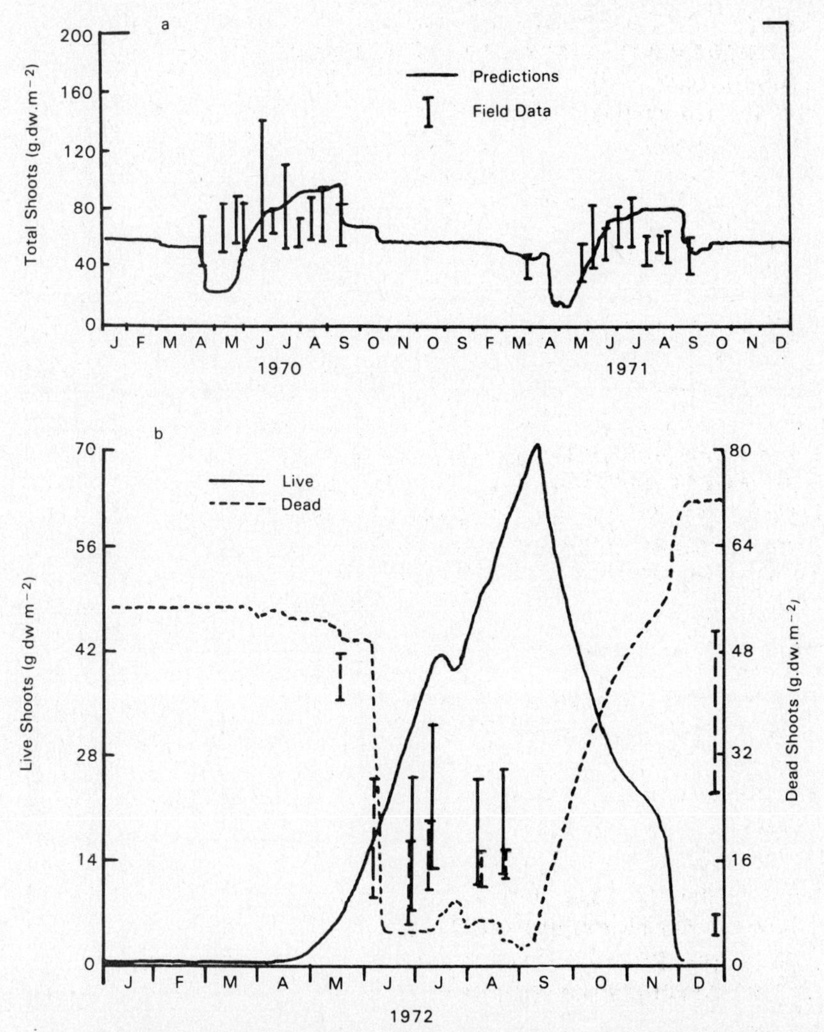

Fig. 69 Comparisons of predictions of the primary production submodel of ELM with field data on warm season grasses at the Pawnee site; a. 1970-71, b. 1972 (after Sauer 1978)

nitrogen, the presence or absence of light or heavy grazing, presence or absence of consumers, the application of extra heat input or the simulation of drought conditions and so forth. The first six experiments covered in figure 70 were run in the field also. The seven remaining experiments were specifically to investigate the properties of the model. Woodmansee presents a set of criticisms and analyses of the ELM model as a whole, pointing out where the product of the modelling effort fell short

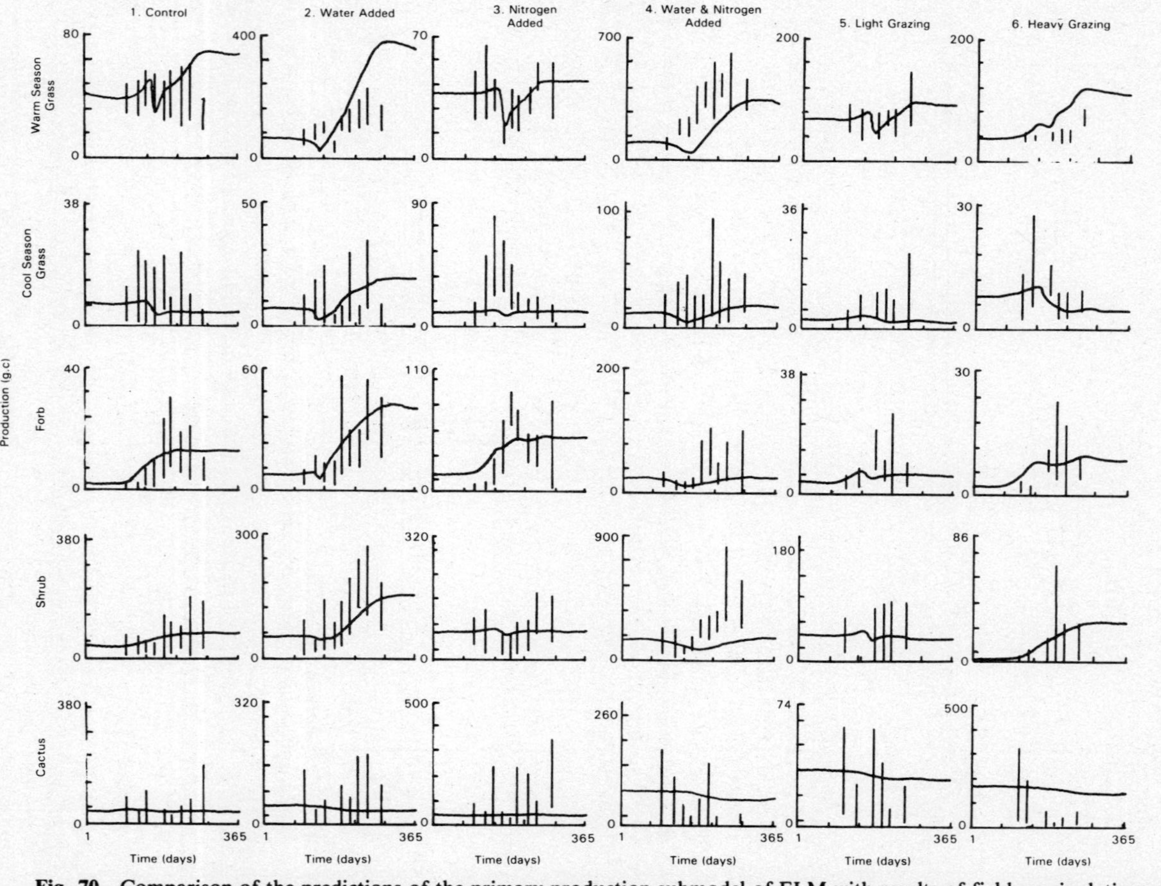

Fig. 70 Comparison of the predictions of the primary production submodel of ELM with results of field manipulations of pastures shown as changes in the standing crop over time (after Sauer 1978)

of the initial objectives laid down by the IBP team. I paraphrase his article in concluding this account of the model.

Woodmansee first pointed out that one of the goals of this modelling effort was to show that the tools of systems analysis, as expressed in simulation modelling, could be applied in the representation of complex ecosystems, leading to a model of the dynamics of grassland in North America. He was able to show that it was indeed possible to build a model of a system as complex as this and to yield useful constructs.

Woodmansee also said that the model represented a collection and integration of information about the grassland ecosystem originating from a great many disparate sources in the literature, in the experience of grassland field workers and following additional observations made by the IBP team. The formulations and parameter values represented the best information available about the ecosystem and the ecological processes involved at the time of modelling and, as such, the whole represented an extremely valuable compilation relating to the state of the art at the time.

Woodmansee identified a function of the model as a communication device, in as much as the ELM model represented a coming together of a great many different pieces of information originating from a variety of disciplines, from biology through meteorology to soil science. The modelling effort and the formulation of the dynamics of the processes involved was an interdisciplinary study in the true sense of the word, and the common format provided by the aims of the programme permitted more than the usual amount of communication by the participants from the various disciplines.

As a research tool, the model is a valuable integrator of information; it is a means of studying and, indeed, simply identifying poorly understood parts of the grassland ecosystem, such as the below-ground system. As mentioned earlier, one of the most valuable things a complex model can do is to identify what you do not know: this can then be used as a guide to future investigations in the field. The model can also be used as a way of testing hypotheses about the dynamics of the grassland ecosystem: Woodmansee spoke of work being done on the effects of grazing, fire and pesticides on the responses and performance of grassland ecosystems. He pointed out that as a generator of hypotheses, the ELM model and its functioning can fit into the general model of scientific investigation: that is, the model can be used to generate output under a variety of given circumstances; these results can then be used as hypotheses and, in the normal scientific manner, field and experimental work can be carried out in an attempt to falsify the hypotheses. As Woodmansee says, this approach requires that careful and diligent attention be given to the parameter values used in the model but, in so far as virtually all the parameters can be ascribed ecological or other biological meaning, this is by no means an unattainable goal. Lastly, the model may be used as a predictive device; indeed, it is the results of this function that are seen in figure 70. As a predictive device the model has an immense potential in management science. The author sees the ELM model as being in the developmental stage still; for the moment, comparisons of experimental results and predicted results obtained through running the model are most valuable as a means of identifying shortcomings in the model itself, which can lead to an improvement in its structure. Ultimately, however, the model and others like it will allow the evaluation of management options by simulation rather than the much more expensive "suck it and see" methods frequently applied in evaluating different strategies for the conservation, exploitation or upgrading of ecosystems.

8.3 The Florida Everglades: An "Energese" Model of an Aquatic System

8.3.1 Background Information

The third and last example of an ecosystem model that I shall examine in this chapter was built using the approach and philosophy of H.T. Odum, for implementation upon an analogue computer. I discussed in chapter 2 the method of systems diagramming proposed by Odum and some of the advantages and disadvantages in its use. Stemming from the work of Odum and his students and colleagues at the University of North Carolina at Chapel Hill and subsequently at the University of Florida at Gainesville, a relatively large number of models have been built, employing what is recognizable as a common set of techniques associated with their analogue and differential nature. A recent volume by Hall and Day (1977) presents a number of these models and others are scattered in the ecological literature. I have chosen to present here a simulation of part of the vast freshwater marshland called the Everglades, an area of some considerable interest from the point of view of conservation and land use in southern Florida. This selection is not an arbitrary one: I make it because the model concerned is relatively simple and, most important of all, is written up in a particularly clear and complete form by Bayley and Odum (1976). My account is based wholly on this paper. Once again I shall use the common subheadings employed in 8.1 and 8.2.

8.3.2 Problem Definition

The predominant vegetation type within the Florida Everglades is sawgrass (*Cladium jamaicense*) marshes, which Bayley and Odum describe as making up 65–70 per cent of the whole region. The production and nutrient dynamics of the sawgrass is intimately tied up with the **water** regime in the marsh and this is the result of run-off from Lake Okeechobee together with incident rainfall. Drainage activities and flood alleviation measures have resulted in restriction in the amount of run-off from the lake which reaches the marsh and have led overall to a much drier environment, to which the marsh communities have responded in a variety of ways. Also of particular significance in this ecosystem is the role of **fires**, which mobilize nutrient and destroy standing vegetation. Fires coincide with drier periods but can also be used in a regulated fashion as a management tool.

These natural processes mentioned above are synthesized in Bayley and Odum's model as a means of exploring the various management options open to the manipulators of the marsh, through the control of water flow and fire. They proposed to use the model to manipulate these and other variables and follow the consequent changes in a variety of state variables. They hoped to identify critical state variables and processes affecting their dynamics and, overall, put together a tool useful for the assessment of many different aspects of the ecosystem.

8.3.3 Systems Identification

Bayley and Odum identified four basic state variables in their model, which represented stocks of energy or material within the marsh ecosystem: **water, phosphorus, peat** and **sawgrass.** The interconnections between these variables and the processes affecting them are illustrated in figure 71, using the techniques of systems diagramming proposed by Odum. It will be seen

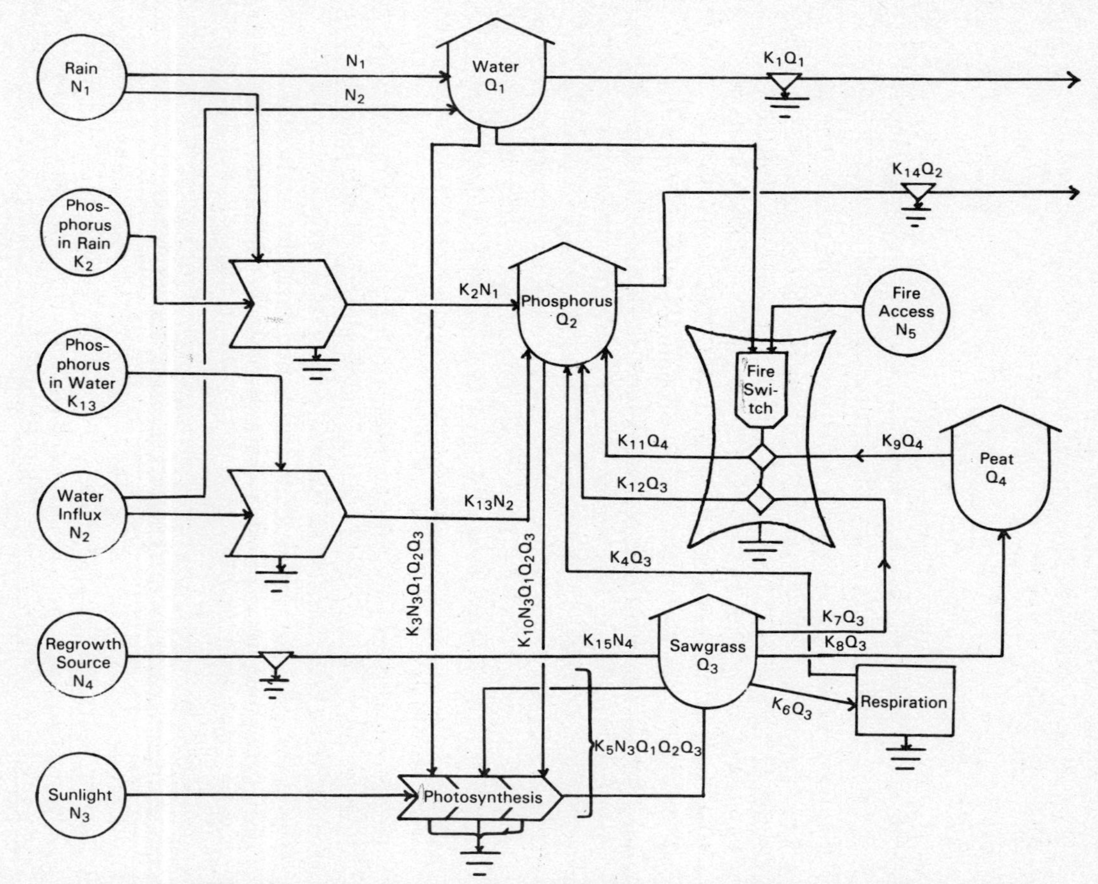

Fig. 71 Systems diagram of the sawgrass marsh modelled by Bayley and Odum. The symbology used is the "energese" of H.T. Odum, explained in chapter 2 (after Bayley and Odum 1976)

from this diagram that, in addition to the four state variables, the authors identify seven driving inputs or variables reflecting incident weather conditions, sunlight, water influx and the source of plant regrowth following a major destruction of the sawgrass component. In addition the **fire "access"**, as it is described, is a controlling variable which operates in the model when the water level falls below a critical point.

8.3.4 Decisions on Model Type

The model built by Bayley and Odum was of the type that has come to be generally associated with that school of systems ecology begun and led by H.T. Odum. These models are designed specifically to be implemented on the analogue computer, hence their type is entirely prescribed by the nature of the computer that is to be used. That is to say, the models are based on sets of differential equations and accordingly are continuous and deterministic in nature. The advantage of a model of this type implemented on an analogue computer is that the interaction between the model and the modeller is achieved in a particularly convenient manner, the magnitudes of the input variables being controlled by the setting of potentiometers — knobs on the front of the computer. Therefore, an investigation of the sensitivity and response of the model to a particular input variable can be achieved simply by turning the knob of the potentiometer. Output is usually graphical, displayed on a cathode-ray screen, and changes in the output can be followed very closely by observation of this graphical output as it responds to changes in potentiometer settings.

8.3.5 Mathematical Formulation

The four differential equations which represent the dynamics of the four state variables identified in the sawgrass ecosystem are shown in table 53. Each contains additive and subtractive components much as the equations of change for the tree-hole model described in 8.1. Each input and output within the differential equations is modified by a constant coefficient represented in the equations by the Ks. Driving variables are represented by the Ns and inputs and outputs to and from the other state variables are represented by the Qs. The equation relating changes in the amounts of **water** present contains inputs from rain and inundation and outputs to photosynthesis and by evapotranspiration.

Table 53. Transition equations for the sawgrass marsh model (modified slightly from Bayley and Odum 1976)

Water Dynamics

$$\frac{dQ_1}{dt} = N_1 + N_2 - K_3 N_3 Q_1 Q_2 Q_3 - K_1 Q_1$$

where		
	N_1	– Input from rain
	N_2	– Input from water influx
	$K_3 N_3 Q_1 Q_2 Q_3$	– Output to photosynthesis
	$K_1 Q_1$	– Loss by evaporation, transpiration and run-off

Phosphorus Dynamics

$$\frac{dQ_2}{dt} = K_2 N_1 + K_{13} N_2 + K_4 Q_3 - K_{10} N_3 Q_1 Q_2 Q_3(+ K_{11} Q_4 + K_{12} Q_3) -(K_{14} Q_2)$$

where		
	$K_2 N_1$	– Input from rain
	$K_{13} N_2$	– Input from water influx
	$K_4 Q_3$	– Input from sawgrass respiration
	$K_{10} N_3 Q_1 Q_2 Q_3$	– Output to photosynthesis
	$K_{11} Q_4$	– Input from peat during fire only
	$K_{12} Q_3$	– Input from sawgrass during fire only
	$K_{14} Q_2$	– Overflow loss

Sawgrass dynamics

$$\frac{dQ_3}{dt} = K_5 N_3 Q_1 Q_2 Q_3 - K_8 Q_3 - K_6 Q_3 + K_{15} N_4 - (K_7 Q_3)$$

where		
	$K_5 N_3 Q_1 Q_2 Q_3$	– Input from photosynthesis
	$K_8 Q_3$	– Loss by death to peat
	$K_6 Q_3$	– Loss to respiration
	$K_{15} N_4$	– Gain from regrowth source
	$K_7 Q_3$	– Loss to fire

Peat Dynamics

$$\frac{dQ_4}{dt} = K_8 Q_3 - (K_g Q_4)$$

where		
	$K_8 Q_3$	– Gain from sawgrass
	$K_9 Q_4$	– Loss to fire

Note: refer also to figure 71 for definition of symbols. Terms in parentheses do not operate all the time.

The equation for **phosphorus** contains inputs from rain and inundation, an input from the respiration of the sawgrass component, and an output to photosynthesis. There are two additional inputs from peat and sawgrass respectively which only operate during fire. In addition, there is an overflow term which operates when the system is saturated with phosphorus. The equation for the dynamics of the **sawgrass** component contains inputs from photosynthesis and a relatively small input from the regrowth source, together with outputs to the peat component following death, to respiration and, under certain circumstances, to fire. The input from the regrowth source needs a little further explanation. Following fire the sawgrass component may be reduced virtually to zero; in real life this is overcome by the small amounts of plant material which may enter the sytem from outside in a variety of ways, either as roots which may shoot or buried seeds that are not affected by the fire. These mechanisms are substituted in the model by the driving variable, **regrowth**. The last equation, governing **peat** dynamics, contains a basic input from sawgrass and a loss by fire from time to time.

8.3.6 Decisions on Computing Methods and Programming

The decisions on computing methods are not in question in this particular type of model, which is wholly designed for a particular implementation using analogue computers. The programming steps are outlined in some detail in the paper by Bayley and Odum, and involve working out scaling factors for each of the parameters in the equations of change, so that full scale deflection on a particular potentiometer in the analogue computer will be such as to produce the full range of values required of a particular component in the model. The actual patching diagram by which the model is transferred directly to the analogue computer is given in full, using standard symbols by Bayley and Odum.

8.3.7 Parameter Estimation

As in the ELM model, Bayley and Odum used a variety of sources to obtain realistic values for the parameters in their model. They were able to obtain some of these from published material, others as a result of personal communication with research workers and, in a few cases, they were reduced to putting in values of convenience where no data were available. Table 54 shows the parameter values used by these authors and their sources for each of the fifteen transfer coefficients involved.

8.3.8 Validation and Experimentation

Bayley and Odum (1976) combine their validation and experimentation phases in the write-up of the sawgrass ecosystem

Table 54. The transfer coefficients, parameter values and their sources for the sawgrass marsh model (from Bayley and Odum 1976)

Coefficient	Explanation	Value Used	Source
K_1	Rate of evaporation of water	0.41 mm/mo.	Clayton and Neller (1938)
K_2	Rate of phosphorus intake from rain	10^{-5}gP/litre	Schneider and Little (1968)
K_3	Rate of sawgrass evapotranspiration	$1.01 \times 10^{-6} m^6 g^{-2} kcal^{-1}$	Clayton and Neller (1938)
K_4	Rate of phosphorus input from sawgrass respiration	8.9×10^{-7}/mo.	Miller (1918)
K_5	Rate of photosynthesis of sawgrass	$3.1 \times 10^{-7} m^3 10^3 g^{-1} kcal^{-1}$	Davis (1946)
K_6	Rate of sawgrass respiration	0.0089/mo.	Miller (1918)
K_7	Rate of loss of sawgrass to fire	3.0 g/mo.	*pers. comm.*
K_8	Rate of sawgrass loss to peat	1.5×10^{-3}/mo.	Davis (1946)
K_9	Rate of peat loss to fire	0.1/mo.	Miller (1918)
K_{10}	Rater of phosphorus loss to photosynthesis	$.0031 \times 10^{-8} m^3 g^{-1} kcal$	Miller (1918)
K_{11}	Rate of phosphorus input from peat by fire	1×10^{-5}/mo.	Miller (1918)
K_{12}	Rate of phosphorus input from sawgrass by fire	3.0×10^{-4}/mo.	Miller (1918)
K_{13}	Rate of P intake from water by influx	10^{-3}mg/litre	Sullivan et al. (1971)
K_{14}	Rate of P loss to overflow	$(K_{14}Q_2$	Internally computed)
K_{15}	Rate of sawgrass input from regrowth source	$(K_{15}N_4$ + 1 g/m^2/mo.	Internally computed)

Source references: Clayton, B.S. and Neller, J.R. (1938) Water control investigations. *Univ. Fla. exp. Stn., Annual Report.*

Davis, J.H. (1946) The peat deposits of Florida. *Fla. geol. Survey, Geol. Bull.* 30: 1-247.

Miller, C.R. (1918) Inorganic composition of peat and of the plant from which it was formed. *J. Agric. Res.* 13: 605-9.

Schneider, R.F. and Little, J.A. (1968) *Characterization of bottom sediments and selected nitrogen and phosphorus sources in Lake Apopka, Florida.* USDA Federal Water Pollution Control Administration, S.E. Water Laboratory, Athens, Ga. 42 pp.

Sullivan, J.H., Fox, J.L., Furman, T.D., Lackey, J.B., Singley, J.E. and Patterson, A.D. (1971) *Water Quality Survey – The Everglades National Park* Water Air Resources Inc. Gainsville, Florida 59 pp.

model. They present output from the model under a variety of conditions of fire regimes, water influx, transpiration levels and phosphorus inflows. The output from some of these runs is presented in figure 72. From these and similar outputs the authors conclude that the vegetation characteristics of the system control the hydrological aspects. In addition, the simple

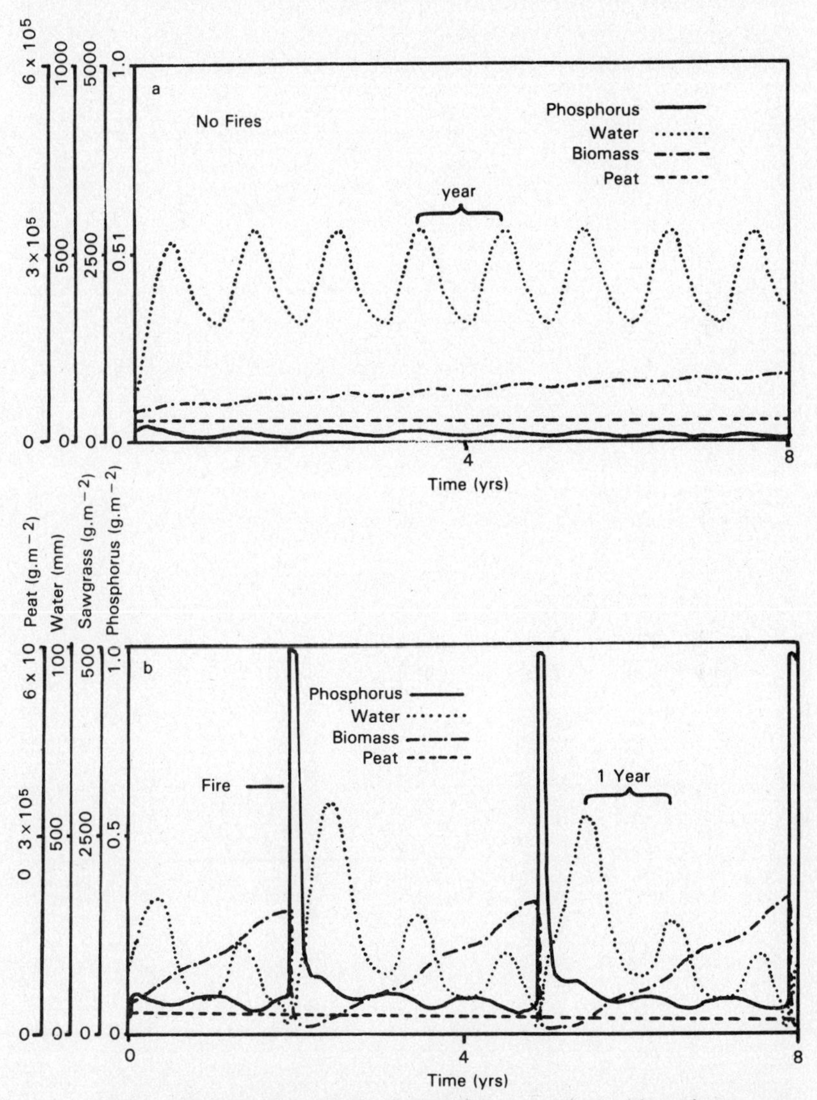

Fig. 72 Outputs from Bayley and Odum's Everglades model: a. with moderate water inflow, low phosphorus levels and no fires – possibly similar to the original state of the Everglades; b. with high water inflow, high phosphorus levels and periodic fires (after Bayley and Odum 1976)

nature of the ecosystem is shown to be the reason for its sensitivity and unpredictability. They conclude that these features coincide with observations that have been made on the Everglades themselves and that much more data is needed for full evaluation of the ecosystem.

Like the ELM model, this one serves a variety of purposes. It is a synthesis of information, a research tool and, potentially at least, a management tool. It identifies what we do not know, by formalizing what we do know about an ecological system.

9

Prospects and Pitfalls

In concluding this account of systems ecology, I intend to indulge myself in that most dangerous of pastimes, prophecy. In any area of science at any point in time, it is possible to make predictions about where major developments are likely to occur in the near future. However, like all prophecies, such predictions depend very much on a personal interpretation of the current state of things. Ecologists in particular have turned to prophecy, from time to time, in their attempts to comment upon the human situation in the twentieth century. I hasten to say that this is not my intention: I shall be restricting my predictions to the strictly scientific aspects of my subject matter. There are three areas in which some trends are apparent in the literature at the moment: the changing role of systems modelling in the theory and application of the science of ecology as a whole, the mathematical and computational developments which seem likely to receive much more attention in the future and, lastly, the prospects in the development of computer technology which will have an important feedback upon the *modus operandi* of the practising systems ecologist. I shall conclude the chapter by outlining some of the pitfalls which await the unwary in systems ecology.

9.1 Prospects

In my estimation the prospects for the development of systems approaches in ecology are very good. Indeed I see the adoption of the philosophy as well-nigh inevitable. As I have pointed out in earlier chapters, the most important development in the subject to date, and one which must form the basis for further development, is the integration of the technology of systems analysis with the skills and intuitions which are often associated with more traditional ecology. That is to say, systems ecologists must be biologists if they wish to make any significant statement on biological systems. In this regard there is a change in the way ecologists are trained, and indeed this work is intended to be effective in this area. It is no longer sufficient for trainee ecologists to concentrate their efforts wholly on learning the biological aspects of the subject, perhaps with a little practical statistics thrown in. Instead, it is vital that such trainees be given a thorough grounding in aspects of mathematics and computing, so much so in fact that we might regard ecology not wholly as a subscience of biology but as an interdiscipline falling between biology and mathematics.

9.1.1 The Role of Systems Modelling in Ecology

The ways in which systems modelling has been used in ecology to date suggest two major roles for the approach in the future: as tools leading towards the production of **generalizations** about natural and man-affected ecological systems, and as means for the more efficient **management** of ecological systems. The time is ripe for a reappraisal of both these roles, and my comments here can do no more than foreshadow such reappraisal. Before considering each of these developing roles in turn, it must be pointed out that the intrinsic value of models as synthesizing and integrating devices cannot be understated. Time and again, the examples in this book have shown models as mechanisms by which the worker is able to identify what he does not know, by organizing and systematizing what he does know within the framework of a conceptual model of a particular system.

Ecological Theory-Building

The formulation of **generalizations** about ecological systems has been the goal of so-called theoretical ecologists for decades, receiving special impetus recently with widening familiarity with mathematical and other technical aids (see, for example, May 1976). To speak of the future in this area requires some consideration of the likely nature of ecological "theory". Clark, Kitching and Geier (1980) treat this somewhat vexed question in the light of some current statements about the philosophy of similarly defined sciences such as geography. These authors suggest that, as a synthetic discipline, ecology may generate principles, precepts, general models and even paradigms, but NOT laws or general theories such as may be anticipated in the primary, reductionist disciplines such as physics or chemistry.

To illustrate the difference here implied, and without getting sidetracked into semantics, I quote two definitions from the Oxford English Dictionary: for precept, "a maxim, a general truth drawn from science", and for (natural) Law, a "correct statement of invariable sequence between specified conditions and specified phenomena". The different degrees of rigour implicit in these two definitions illustrate the perceived distinction between the **secondary** and **primary** sciences. Thus ethology and evolutionary biology are seen as **primary** sciences of great relevance to ecology within which general laws, the laws of neodarwinian evolution, may be formulated. The recent flowering of subdisciplines such as sociobiology and population genetics are evidence of this process, which must bear directly on the ecologist's worldview and scientific approach.

The distinction between these primary and secondary sciences reflects a difference in the objects that each has as its subject matter. For the primary sciences these are **primary objects** — discrete, recognizable entities such as individual molecules, cells, organs and even organisms. The **secondary** sciences study, not such primary objects, but so-called **systemic aggregates** or, simply, systems as defined in this book. Parenthetically, the logical consequences of this recognition is that all of ecology becomes **systems ecology** in the present sense!

Having covered this philosophical ground, the role for modelling in non-applied ecology becomes quite clear. Models of ecological processes, such as those of predation and movement discussed in chapter 7, should generate principles about the structure and functioning of the processes. Ultimately one can imagine a set of such models available to the ecologist like a set of building blocks for use in higher level studies. As general principles, these will be approximations requiring modifications or additions for application to particular situations, but they will be "truths" at the level of "precept" discussed above. It should be said that process models feed not only into ecological endeavour but also into ethology and evolutionary biology, in so far as the interpretation of the behaviour of individual organisms is involved, drawing upon and adding to the tenets of evolutionary theory. In this region, any particular worker cannot and must not purport to be wholly an ecologist, an ethologist, or whatever: he is involved in interdisciplinary study of a particular set of phenomena. His results, however, may be picked up by other workers in more specific disciplines.

Population and other higher level models play two roles in the scientific process, one of which will be considered briefly in the discussion of models in management below. The other is parallel to the role of process models; that is, in time they will form a set of **case studies**, key works against which any newly considered system can be matched. The trend towards this situation becomes evident when one contemplates series of studies of the life systems of related species, such as the aphids described in chapter 6. Although constructed as independent exercises, the common features of these models, further shared with those of other insect life systems, support the idea that there is a canonical pattern which distils the characteristics out of all the specific case histories. Of course the way of proceeding to such a distillation is not entirely clear at this stage. Certainly an on-going necessity is the overview by competent ecologists of the whole range of specific models, to encourage the perception of commonalities as and when they appear. In other words, to recapitulate earlier comments, it is vital that training schemes within the science lead to enhanced competence in the apprecia-tion and evaluation of the whole range of systems models being produced.

Management of Ecological Systems

The provision of an environmental impact statement, the deter-mination of proper fishing practices, the development of ways of designing national parks so as to preserve the integrity and diversity of the natural communities they contain, are all real world problems, in the solution of which ecological principles must play a large part. These and a wide range of comparable problems demand that we predict the results of manipulations of very complex entities — ecosystems, populations and natural communities — involving many components and processes. We know, intuitively, that our understanding and prescriptions for the management of these entities must come to grips with this complexity.

All of these and other considerations underline the point, made early in this book, that the construction of complex systems models to tackle such problems is the obvious, and fre-quently the only, way to proceed to their solution.

Many of the models that have been examined in detail have a basic economic *raison d'être*. Gutierrez's aphid model, the ELM grassland model, and Bayley and Odum's model of the Everglades are cases in point. However, the management role of these works although foreshadowed, was not realized extensive-ly in practice. In this regard, the workers concerned have reflected simply the state of their art at the time. Each of these, and many other models of the sixties and early seventies, were involved both in the process of investigating ecological phenomena and in developing the methodology and technology of modelling, such that, subsequently, models directed specifically at ecological management could be constructed and used. These earlier labours have recently borne fruit, and a wide range of modelling applications have appeared. I shall note some of these briefly here, giving key and interesting references, to work in each of the basic areas of population management.

Pest management. The development of strategies for the control of pests, particularly insects, took early advantage of systems techniques, perhaps because many of the earliest simulation models were built of insect populations. In this area of modelling, the concept of the life system of Clark *et al.* (1968) has played a major part. As early as 1973, the compendium by Geier *et al.* reflected several of these early ventures in applied systems analysis. Within that volume specific models of the dynamics of algal flies (Wiegert), cereal leaf beetle (Haynes and Barr) and pine bark beetles (Stark) are presented, together with a useful general treatment on modelling of pest situations by Conway. Most of these models were built, overtly or covertly, on the life system idea and, indeed, in a slightly later review, Ruesink (1976) identifies the life system as "... a useful point of departure for applied entomologists". By 1976 Ruesink was able to identify well over a hundred simulation models for pest management, ranging from those that examined particular species, through those that examined a particular class of control methods to a very few "general" models for pest control.

No doubt, in 1982, we could multiply Ruesink's examples several-fold. Suffice it to say, now, that models of most methods of pest control are available. These include simple insecticidal control (Watt 1961), biological control by natural enemies (a large range of specific models including Hughes and Gilbert 1968, Hassell and Varley 1969, Fransz 1974), sterile male release methods (for example, Berryman 1967) and the use of chemosteritants (for example, Barclay 1981). In addition, a body of models examines major pest species allowing a range of control options (for example, Stinner *et al.* 1974, on *Heliothis*; Weidhaas 1974, on mosquitoes, Harmsen *et al.* 1974, on cutworm; Walters and Peterman 1974, on spruce budworm; Sasaba and Kiritani 1975, on rice leaf-hoppers; Mitsch 1976, on water hyacinth; and others already mentioned). Yet other work examines the complex of pests on a particular crop (for example,

Blood *et al.* 1975 and Gutierrez *et al.* 1975, on cotton; and Botkin *et al.* 1972, on forests). Lastly, note must be taken of the seminal work of Shoemaker (1973a,b,c), who constructed general, open-ended models of pest control, including both biological and economic information.

The development of so-called **on-line systems models** also shows great promise in this area. Basically, these are models which can be used directly by the manager — farmer, forester or pest controller — to obtain predictions about likely pest outbreaks in his crop. Using terminals hooked to computers at state or national Departments of Agriculture or similar institutions, users are able to feed into such models current information on local climate or other environmental variables and results of "scouting" surveys for key pest species; they obtain, in return, the model's predictions for their area, sometimes with specific recommendations about appropriate control measures. Among the pioneers in this area were a Michigan-based group of workers and their efforts, together with an account of the rationale and operation of such systems, are described by Croft *et al.* (1976). The on-line facilities developed by these workers include models of apple and asparagus pests. A second group of workers based in Indiana developed comparable models for the management of alfalfa pests (for example, Huber *et al.* 1973). Room (1979) has developed on-line models for the management of pests of cotton in Australia.

A complementary aspect of such on-line models is presented by Welch and Croft (1979), who discuss the data collection process vital for proper pest management. With reference specifically to red spider mite outbreaks, they examine the biological, statistical and economic components which enter into decision making vis-à-vis control and the use of simulation models in collating incoming data and underpinning subsequent management operations.

Fisheries management. Beverton and Holt (1957), as noted in chapter 1, summarized and extended the use of mathematical models in fisheries. Given this predisposition to the use of such methods, it is not surprising that fisheries scientists were among the earliest to exploit simulation methods in their work. 1969 saw a special issue of the Transactions of the American Fisheries Society devoted to the use of simulation models, which contained seminal articles by Walters, Silliman, Patten and Paulik aimed at drawing the attention of the appropriate professionals to the power of these techniques.

Subsequently, specific models have been published from time to time. Among the more notable of these are Larkin's (1971) study of sockeye salmon, Grant and Griffin (1979) on shrimp in the Gulf of Mexico, Tillman and Stadelman (1976) on the anchovy fishery and Southward (1968) on Pacific halibut. Specific modelling techniques applied within fisheries models include Leslie matrices (Horst 1977, see also Usher 1972, 1976), catastrophe theory (Jones and Walters 1976), dynamic optimization (Walters and Hilborn 1976) and stochastic dynamic programming (Hilborn 1976). In all of these developments, the dominant role of fisheries scientists under the direction and encouragement of C.J. Walters at the University of British Columbia is evident, and members of this group remain the leaders in this field.

Most recently, Larkin and Gazey (1981) have pointed to one direction that is proving rewarding and holds considerable promise for the future. This is the development of models which represent and which can be used to help manage whole fisheries involving many species. These authors refer to such developments for the eastern Bering Sea (Laevestu and Favorite 1977) and the North Sea (Andersen and Ursin 1977), and foreshadow development of such models for tropical seas. As is implied by Larkin and Gazey, simulation has become an established part of fisheries science; its role seems assured in the development of new and more complex fisheries, as well as in the restoration and maintenance of those already established.

Game management. The problems facing the managers of stocks of terrestrial mammals and birds which are exploited by hunting have many parallels with those addressed by managers of fisheries. Their complex, multivariate nature suggests that an approach to problem solving through simulation modelling would be profitable. Early indications were that such a development would occur. Models such as that of Walters and Bandy (1972) set the pace, and specific models have appeared from time to time (see, for example, Walters *et al.* 1975, on barren ground caribou; Sinclair 1973, on African buffalo; Brown *et al.* 1976, on North American mallard; and Roseberry 1979, on bobwhite quail). However, overall, I do not think the early promise has been exploited; the role of simulation modelling seems to be rather less favoured now than five years ago. The demonstrable success of such techniques in pest and fisheries management suggests that this may be a case of resistance by a conservative market, rather than a lack of applicability.

9.1.2 Some Mathematical Prospects

I have mentioned a variety of different mathematical techniques which have been used to date to a greater or lesser extent in systems ecology, and I have suggested some areas in which a great expansion in the use and application of techniques may occur. In this concluding account I wish to highlight three of the areas already mentioned and to foreshadow the introduction of other techniques which may occur in the near future.

1. The application of **matrix theory** from the realm of **linear algebra** is one method in which I see a great expansion in application in the modelling of natural systems. Chapter 6 examined the nature of Leslie matrices both in the basic form, as proposed in 1945, and in variously modified forms introduced by authors such as Williamson, Usher, Enright and Ogden. The use of matrices to represent spatial heterogeneity, different behavioural states in animals, and in a variety of other applications has already been suggested in the literature or is likely to be developed in the near future. The mathematical succinctness of matrices and their great flexibility both mathematically and in the manner in which they can be fitted into larger, more detailed, simulation models, suggests that they are an ideal mathematical tool for population models and, indeed, have applications in models at other levels of resolution as well. They are likely to have special application in the development of models for management, especially for animal populations, in so far as they maintain the age-specific identity of the population, an all important feature in the development of cropping and control strategies.
2. There is currently considerable interest within systems modelling in **catastrophe theory**, a subject only briefly touched on so far. Catastrophe theory is a recent development from the field of algebra known as **differential topology**. This field of mathematics is particularly difficult for all but a few specialized mathematicians but, recently, the essence of the ideas involved has penetrated into ecology. Very simply, catastrophe theory provides a series of functional forms, which can be fitted to surfaces which relate the changes in value of some key ecological variables, such as population number, to one or more other variables. These surfaces include in themselves discontinuities, folds by analogy, which could make them particularly useful for representation of many ecological

and other biological situations. The most extensively documented example is that of Jones (1977), relating changes in levels of the spruce bud-worm moth, a pest of coniferous forests in North America, to a small set of environmental variables. Jones claims to show that one of the simpler models of catastrophe theory can be used to predict, and hence forestall, the levels of abundance at which the population would "explode" to cause a massive outbreak so characteristic of the species. The important thing about this approach is that it does not require detailed knowledge of the functioning of the life system of the subject species. It is essentially a fitting procedure akin to the least squares method of statistical regression, and can be applied given even a relatively restricted data base. It requires far less effort in analysis and data collection than the preparation of a full life system model of the type described earlier. In this respect it is interesting to speculate that perhaps it provides the tool to match the situation identified by Geier and Clark (1979), where they point out the uneconomical and impractical aspects of building full scale life systems models for every pest situation facing the ecologist. The application of the techniques of catastrophe theory remains in its infancy in ecology and, indeed, some authors have suggested that it may be an inappropriate tool outside of the narrow mathematical bounds set down for it by its originators. However, it is certainly one area with a great deal of promise, which should see many exciting developments in the near future.

3. Another area of proven worth and great potential in systems ecology and management has developed in parallel with the application of catastrophe theory, when considering the spruce bud-worm system of North America (Holling and Dantzig 1977, Peterman 1977). The dynamics and management of the pest were examined using optimization methods of dynamic programming derived from the theory of **operations research**. Essentially, this involves finding an optimal solution to a particular question, contingent upon whatever sequence of events has occurred previous to the point of analysis. Once again, what is involved is an analysis of the response surface of a relatively simple statistical model of the problem situation, and it holds great promise for pest management procedures, where there is likely to be an optimal strategy which will maximize the chances of effective control.

Having looked at areas of mathematics employed in the application of systems modelling techniques in particular situations, I will move on now to a much more speculative area. Essentially, we must consider developments in the **reflective study** of the models of ecological situations that have been built, rather than ways in which further models might be built. There are a few workers who are attempting to look over the whole field of quantitative ecology to identify trends and common features across models that have been built, and endeavouring to draw conclusions about the different techniques involved and, indeed, about the subject systems themselves. Much of this work remains highly speculative and is unpublished. However, progress must be made in this direction if we are not to be submerged ultimately by a deluge of unconnected and unrelated models of a multiplicity of systems. One particularly interesting development in this field is by Niven (1980), in which she takes the first tentative steps towards defining a "calculus" for population studies. Niven argues that the behaviour of organisms in the population situation is so different from the behaviour of continuously varying entities, the ideal subject matter of the normal differential calculus, that it is inappropriate to attempt to apply such mathematics to biological populations. She argues that it should be possible to define a mathematics specifically for this purpose, and in her first paper on the subject (1980) has taken the initial steps towards the definition of such a calculus. Subsequently, in a series of unpublished working papers, Niven has defined the environment of selected species of animals based on her calculus and a heuristic extension of it — the **envirogram**. Figure 73 is an example of one such envirogram, for the water snail, *Limnaea peregra.* It remains to be seen whether or not this general approach will prove as revolutionary as it appears, but it is certainly a first brave attempt to obtain a true mathematical overview of the dynamics of populations, without constraining those dynamics so that they can be represented by the conventional mathematics of differential or difference equations.

9.1.3 Some Computing Prospects

The speed and dramatic nature of technological developments within the computing field are extremely well-known and anything that I can say here is likely to be out of date within a very short time. One or two general trends that can be identified are in some ways already here. The first of these is the general

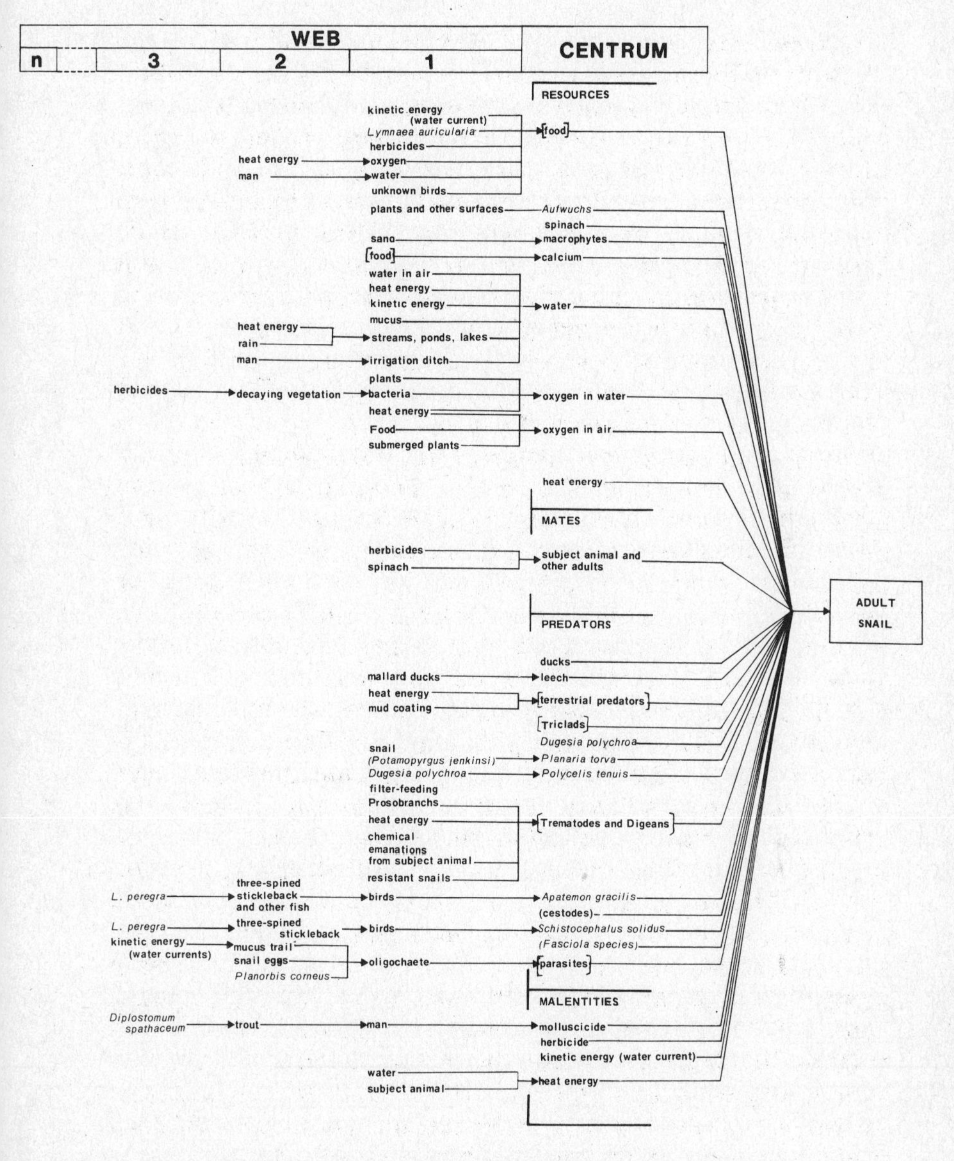

Fig. 73 Envirogram of the water snail, *Limnaea peregra* (reproduced, by permission, from an unpublished manuscript by Dr B.S. Niven)

availability of computers. There has been an exponential increase over the last ten years in the availability of computers, and more recently a trend away from the monolithic giant computer, held in some central facility, towards the much more readily available and easily used desk-top micromachine, which is treated almost as a glorified calculating machine. Such small, cheap and highly accessible machines will become, inevitably, basic tools in systems ecology and, indeed, in many other areas of computer application in the very near future. The major, centrally installed computers will then be reserved either for very large jobs or will be involved in complex time-sharing procedures, such that they act almost with the ease of a desk-top machine for a particular user.

At the moment, particular computer installations have special features which are not shared with all other installations. One centre may have a particular language compiler, for example, not shared by its neighbours. The same argument may be true of special packages for statistical or other sorts of analysis, for special input/output devices and so on. Up until now, the aim of each computer centre could have been considered to be the accumulation of as many as possible of the manifold facilities available, both within the computer and peripheral to it. However, the development of major networks linking computers via ordinary telephone lines has obviated the need for this expansion within each centre. I can foresee the time when the computer will be matched to the job, rather than the job to whatever computer is on hand. Thus, for example, if I have a job which requires the use of the simulation language, DYNAMO, and my local installation does not have an appropriate compiler, I will simply find out where such a compiler is available and address my programme to that particular installation, through the network linking my local facilities to those elsewhere within the country or even internationally.

A further development which may be expected is the increased use of **automatic data collecting devices** to provide the base on which ecological analyses, including model building, can be made. For some years now, devices have been available known as data-loggers, which convert electrical impulses from a great variety of measuring devices in the laboratory or field into codes which can be read, by an appropriate input device, directly into the computer. These have been particularly valuable in the establishment of remote weather stations and similar unmanned facilities. Such automation is likely to come to more mundane laboratory and field situations in the future. Thus, instead of sequences of data points being written down, they may well be fed directly into a computer for processing, either in parallel with the collection process or afterwards. The use of cassette recorders, light pens and other sophisticated input devices will facilitate this sort of data collection.

9.2 Pitfalls

It is highly appropriate to conclude this work with comments upon three pitfalls which have shown themselves to be ready to engulf the unwary modeller in ecology. All three have been encountered in one form or another in the preceding chapters, but by drawing them together I hope to give them additional emphasis.

Time and time again models have been built and presented in the literature without any real identification, either explicit or implicit, of why the model was being built. In other words, the first step in the model building procedure, the problem identification process, has been omitted. This is a common fault, not only in modelling approaches to ecology, but in other sections of science also. It is not sufficient to measure, describe or model something just because it is there. That may be a necessary preliminary step, but in order to understand or manage the systems involved, we must define one or a series of questions which we wish to address to the system, through whatever technology is available. Careful identification of such questions may lead the ecologist to decide that a complex simulation model is not appropriate as a way of seeking answers, but it may lead him to a quite different set of techniques. This may well reduce the number of models built, but it should have the salutary effect of making those that are built more effective as tools in ecological research and application.

Having overcome the aura and mystique which surround the use of the computer and its programming, the naive modeller frequently discovers just how easy it is to put together ideas in the form of simulation models. It is much easier, less time consuming and more comfortable to sit in the laboratory and write programmes than it is to get involved in data collection, either through laboratory experimentation or field work in order to provide the data base necessary for the initial construction and the validation phase of the modelling process. Having discovered this discrepancy, some modellers decide that their role as ecologists is to write complex models of a variety of ecological processes and phenomena, which they present for others to evaluate. They become hypnotized by technology to the eternal detriment of their science. It cannot be overstressed that the array of systems analytical techniques presented and discussed at length in this book is simply one of a variety of techniques and approaches which the modern ecologist must have in hand, in order to do the job as efficiently and effectively as possible. Just as slavish adherence to single approaches has led to sterility in ecology in the past, so a slavish devotion to systems modelling should equally be avoided. The "compleat" ecologist of the eighties and nineties must be able to handle a computer with the same facility as a bomb calorimeter, Tullgren funnel or Secchi disc. Knowledge of how to operate within the great variety of technologies now available must be part of every ecologist's background, as must the knowledge of when to use each of them.

Lastly, ecology is predominantly a subscience of biology. It examines the way in which living organisms interact with each other and the many and varied facets of their environment in natural and artificial circumstances. When we pose ecological questions, we are asking about the biology of particular situations and, ultimately, answers to these questions and the underlying phenomena must lie in a true appreciation of the biology of the situation. The ecologist needs to be an interdisciplinarian in the true sense of the word, able to draw upon aspects of mathematics, computing, meteorology, chemistry, physics and a great many other fields as well as biology, but biology must play the central role. Thus in building sophisticated systems models of natural situations, it is vital to maintain close contact with the biology of the situation being modelled. It is all too easy to build biological nonsense into a model of a system with which one is not intimately familiar as a biologist, and the oft-quoted maxim "garbage in, garbage out" is one of the undeniable universals of systems ecology. There should perhaps be a rule which says that no ecologist should be permitted to build a model of a system through which he has not trudged and which he has not studied under a variety of weather and seasonal conditions.

References

Andersen, K.P. and Ursin, E. 1977. A multispecies extension to the Beverton and Holt theory of fishing, with accounts of phosphorus circulation and primary production. *Meddr. Danm. Fisk. -og Havunders* N.S. **7**: 461-75.

Ashby, W.R. 1956. *An Introduction to Cybernetics*. London: Chapman and Hall.

Ashley, R. 1974. ANS COBOL. New York: John Wiley.

Barclay, H.J. 1981. Population models on the release of chemosterilants for pest control. *J. appl. Ecol.* **18**: 679-95.

Bartee, T.C. 1960. *Digital Computer Fundamentals*. Tokyo: McGraw-Hill.

Baumgaertner, J.U., Frazer, B.D., Gilbert, N., Gill, B., Gutierrez, A.P., Ives, P.M., Nealis, V., Raworth, D.A. and Summers, C.G. 1981. Coccinellids (Coleoptera) and aphids (Homoptera). *Can. Ent.* **113**: 975-1048.

Bayley, S. and Odum, H.T. 1976. Simulation of inter-relationships of the Everglades marsh, peat, fire, water and phosphorus. *Ecol. Modelling* **2**: 169-88.

Berryman, A.A. 1967. Mathematical description of the sterile male principle. *Can. Ent.* **99**: 858-65.

______.1981. *Population Systems — A General Introduction*. New York and London: Plenum.

Bertalanffy, L. von. 1964. *General System Theory: Foundations, Development, Applications*. New York: George Braziller.

Beverton, R.J.H. and Holt, S.J. 1957. *On the Dynamics of Exploited Fish Populations*. Min. Agric. Fish. Food, Great Britain, London, H.M.S.O.

Birch, L.C. 1948. The intrinsic rate of natural increase of an insect population. *J. Anim. Ecol.* **17**: 15-26.

Bliss, L.C., Heal, O.W. and Moore, J.J. 1981. *Tundra Ecosystems: A Comparative Analysis*. Cambridge: Cambridge University Press.

Blood, P.R.B., Longworth, J.W. and Evenson, J.P. 1975. Management of the cotton agroecosystem in southern Queensland: a preliminary modelling framework. *Proc. ecol. Soc. Aust.* **9**: 230-49.

Botkin, D.B., Janek, J.F. and Wallis, J.R. 1972. Some ecological consequences of a computer model of forest growth. *J. Ecol.* **60**: 849-72.

Boyce, W.E. and DiPrima, R.C. 1965. *Elementary Differential Equations and Boundary Value Problems.* New York: John Wiley.

Breymeyer, A.L. and van Dyne, G.M., eds. 1980. *Grasslands, Systems Analysis and Man.* Cambridge: Cambridge University Press.

Brown, G.M., Hammack, J. and Tillman, M.F. 1976. Mallard population dynamics and management models. *J. Wildl. Manag.* **40**: 542-55.

Bulmer, M.G. 1965. *Principles of Statistics.* Edinburgh and London: Oliver and Boyd.

Burch, J.G. (Jr.) and Strater, F.R. 1974. *Information Systems: Theory and Practice.* Santa Barbara: Hamilton Publishing Company.

Calow, P. 1973. The relationship between fecundity, phenology and longevity: a systems approach. *Am. Nat.* **107**: 559-74.

Clark, L.R., Geier, P.W., Hughes, R.D. and Morris, R.F. 1967. *The Ecology of Insect Populations in Theory and Practice.* London: Methuen.

Clark, L.R., Kitching, R.L. and Geier, P.W. 1979. On the scope and value of ecology. *Protection Ecology* **1**: 223-43.

Cole, L.C. 1954. The population consequences of life-history phenomena. *Quart. Rev. Biol.* **29**: 103-37.

Coupland, R.T., ed. 1979. *Grassland Ecosystems of the World: Analysis of Grasslands and their Uses.* Cambridge: Cambridge University Press.

Cox, D.R. and Hinkley, D.V. 1974. *Theoretical Statistics.* London: Chapman and Hall.

Croft, B.A., Howes, J.L. and Welch, S.M. 1976. A computer-based, extension pest management delivery system. *Environ. Entomol.* **5**: 20-34.

Cuff, W.R. and Hardman, J.M. 1980. A development of the Leslie matrix formulation for restructuring and extending an ecosystem model: the infestation of stored wheat by *Sitophilus oryzae. Ecol. Modelling* **9**: 281-305.

Davis, G.B. 1965. *Introduction to Electronic Computers.* New York: McGraw-Hill.

Deevey, E.S. 1947. Life tables for natural populations in animals. *Q. Rev. Biol.* **22**: 283-314.

Dill, L.M. 1973. An avoidance learning submodel for a general predation model. *Oecologia* (Berl.) **13**: 291-312.

Dixon, A.F.G. 1973. *Biology of Aphids.* London: Edward Arnold.

Draper, N.R. and Smith, H. 1966. *Applied Regression Analysis.* New York: John Wiley and Sons.

Dyne, G.M. van and Anway, J.C. 1976. A research program for and the process of building and testing grassland ecosystems models. *J. Range Manag.* **29**: 114-22.

Ehrenfeld, D.W. 1970. *Biological Conservation.* New York: Holt, Reinhart and Winston.

Einarson, A.S. 1945. Some factors affecting ring-necked pheasant population density. *Murrelet* **26**: 39-44.

Elton, C.S. 1966. *The Pattern of Animal Communities.* London: Methuen.

Emmel, T.C. 1976. *Population Biology.* New York: Harper and Row.

Enright, N. and Ogden, J. 1979. Applications of transition matrix models in forest dynamics: *Araucaria* in Papua New Guinea and *Nothofagus* in New Zealand. *Aust. J. Ecol.* **4**: 3-23.

Forrester, J.W. 1961. *Industrial Dynamics.* Cambridge: M.I.T Press.

______.1968. *Principles of Systems.* Cambridge: Wright-Allen Press.

______.1971. *World Dynamics.* Cambridge: Wright-Allen Press.

Fransz, H.G. 1974. *The Functional Response to Prey Density in an Acarine System.* Wageningen: PUDOC.

Frazer, B.D.and Gilbert, N. 1976. Coccinellids and aphids. *J. ent. Soc. Brit. Columbia* **73**: 33-56.

Gadgil, M. 1971. Dispersal: population consequences and evolution. *Ecology* **52**: 253-61.

Gallopin, G.C. 1972. Structural properties of food webs. In *Systems Analysis and Simulation in Ecology*, Vol. II ed. B.C. Patten, pp. 241-82.

Gause, G.F. 1934. *The Struggle for Existence.* New York: Hafner.

Geier, P.W., Clark, L.R., Anderson, D.J. and Nix, H.A., eds. 1973. *Insects: Studies in Population Management.* Mem. ent. Soc. Australia **1**, 295pp.

Geier, P.W. and Clark, L.R. 1976. On the developing use of modelling in insect ecology and pest management. *Aust. J. Ecol.* **1**: 119-27.

______.1979. The nature and future of pest control: production and process or applied ecology. *Prot. Ecol.* **1**: 79-101.

Gilbert, N.E. 1973. *Biometrical Interpretation.* Oxford: Clarendon Press.

Gilbert, N. and Gutierrez, A.P. 1973. A plant-aphid-parasite relationship. *J. Anim. Ecol.* **42**: 323-40.

Gilbert, N., Gutierrez, A.P., Frazer, B.D. and Jones, R.E. 1976. *Ecological Relationships.* Reading: Freeman.

Gilbert, N. and Hughes, R.D. 1971. A model of an aphid population — three adventures. *J. Anim. Ecol.* **40**: 525-34.

Goldberg, S. 1961. *Difference Equations.* New York: John Wiley.

Goodall, D.W. and Perry, R.A., eds. 1979. *Arid Land Ecosystems Structure, Functioning and Management.* Vol. I. Cambridge: Cambridge University Press.

Gordon, G. 1969. *System Simulation.* Englewood Cliffs: Prentice-Hall.

Goult, R.J., Hoskins, R.F., Milner, J.A. and Pratt, M.J. 1973.

Applicable Mathematics: A Course for Scientists and Engineers. London and Basinstoke: Macmillan.

Grant, W.E. and Griffin, W.L. 1979. A bioeconomic model of the Gulf of Mexico Shrimp Fishery. *Trans. Am. Fish. Soc.* **108**: 1–13.

Greig-Smith, P. 1964. *Quantitative Plant Ecology.* 2nd ed. London: Butterworths.

Griffiths, K.J. and Holling, C.S. 1969. A competition submodel for parasites and predators. *Can. Ent.* **101**: 785-818.

Gustafson, J.D. 1978. Appendix 1A, SIMCOMP 3.0. In *Grassland Simulation Model*, ed. G.S. Innis. NY, Heidelberg, Berlin: Springer-Verlag.

Gutierrez, A.P., Falcon, L.A., Loew, W., Leipzig, P., and Bosch, R. 1975. An analysis of cotton production in California: a model for Acala cotton and the effects of defoliators on its yields. *Environ. Entomol.* **4**: 125-36.

Gutierrez, A.P., Havenstein, D.E., Nix, H.A. and Moore, P.A. 1974a. The ecology of *Aphis craccivora* Koch and subterranean clover stunt virus in south-east Australia. II. A model of cowpea aphid populations in temperate pastures. *J. Appl. Ecol.* **11**: 1-20.

______.1974b. The ecology of *Aphis craccivora* Koch and subterranean clover stunt virus in south-east Australia. III. A regional perspective of the phenology and migration of the cowpea aphid. *J. appl. Ecol.* **11**: 21-35.

Gutierrez, A.P., Morgan, D.J. and Havenstein, D.E. 1971. The ecology of *Aphis craccivora* Koch and subterranean clover stunt virus in south-east Australia. I. The phenology of aphid populations and the epidemiology of virus in pastures in south-east Australia. *J. appl. Ecol.* **8**: 699-721.

Hall, C.A.S. and Day, J.W., eds. 1977. *Ecosystem Modelling in Theory and Practice: an Introduction with Case Histories.* New York: John Wiley and Sons.

Hardman, J.M. 1976a. Deterministic and stochastic models simulating the growth of insect populations over a range of temperatures under Malthusian conditions. *Can. Ent.* **108**: 907-24.

______.1976b. Life table data for use in deterministic and stochastic simulation models predicting the growth of insect populations under Malthusian conditions. *Can. Ent.* **108**: 897-906.

______.1978. A logistic model simulating environmental changes associated with the growth of populations of rice weevils, *Sitophilus oryzae*, reared in small cells of wheat. *J. appl. Ecol.* **15**: 65-87.

Hargrave, J. 1972. *Analogue Computing*. Bletchley: Open University Press.

Harmsen, R., Cheng, H.H. and Reid, D.G. 1974. Dynamics of army-worm populations on tobacco crops in south-western Ontario. I.

A preliminary simulation model for the crop-pest system. *Proc. ent. Soc. Ont.* **105**: 80-85.

Harper, J.L. 1977. *Population Biology of Plants.* London, New York and San Francisco: Academic Press.

Hassell, M.P. 1975. Density dependence in single-species populations. *J. Anim. Ecol.* **44**: 283-95.

______.1976. *The Dynamics of Competition and Predation.* London: Edward Arnold.

______.1977. Some practical implications of recent theoretical studies of host-parasitoid interactions. *Proc. XV int. Congr. Ent.*, pp. 608-16.

______.1978. *The Dynamics of Arthropod Predator-Prey Systems.* Princeton: Princeton University Press.

Hassell, M.P. and Varley, G.C. 1969. New inductive model for insect parasites and its bearing on biological control. *Nature* (Lond.) **223**: 1133-37.

Hilborn, R. 1976. Optimal exploitation of multiple stocks by a common fishery: a new methodology. *J. Fish. Res. Bd. Canada* **33**: 1-5.

Hillier, F.S. and Lieberman, G.J. 1967. *Introduction to Operations Research.* San Francisco, Cambridge, London, Amsterdam: Holden-Day Inc.

Holling, C.S. 1959. Some characteristics of simple types of predation and parasitism. *Can. Ent.* **91**: 385-98.

______.1964. The analysis of complex population processes. *Can. Ent.* **96**: 335-47.

______.1965. The functional response of predators to prey density and its role in mimicry and population regulation. *Mem. ent. Soc. Canada* **45**: 1-60.

______.1973. Resilience and stability of ecological systems. *Ann. Rev. Ecol. Syst.* **4**: 1-23.

Holling, C.S. and Dantzig, G.B. 1977. *Determining optimal polices for ecosystems.* Institute of Resource Ecology, University of British Columbia, Working paper R-7-B, 41pp.

Horn, H.S. 1975. Markovian properties of forest succession. In *Ecology and Evolution of Communities,* ed. M.L. Cody and J.M. Diamond. Cambridge and London: Harvard University Press, pp. 96-211.

Horst, T.J. 1977. Use of the Leslie matrix for assessing environmental impact with an example for a fish population. *Trans. Am. Fish. Soc.* **106**: 253-57.

Howe, R.W. 1952. The biology of the rice weevil, *Calandra oryzae* (L). *Ann. appl. Biol.* **39**: 168-80.

Hubbell, S.P. 1971. Of sowbugs and systems: the ecological bioenergetics of a terrestrial isopod. In *Systems Analysis and Simulation in Ecology,* ed. B.C. Patten, Vol. I, pp. 269-324.

Huber, R.J. and Giese, R.L. 1973. The Indiana alfalfa pest management program. *Proc. North. Centr. Branch ent. Soc. Am.* **28:** 139–43.

Hughes, R.D. 1963. Population dynamics of the cabbage aphid, *Brevicoryne brassicae* (L). *J. Anim. Ecol.* **32**: 393–424.

Hughes, R.D. and Gilbert, N. 1968. A model of an aphid population — a general statement. *J. Anim. Ecol.* **37**: 553-64.

Innis, G.S. 1975. Role of total systems models in the grassland biome study. In *Systems Analysis and Simulation in Ecology,* ed. B.C. Patten, Vol. III, pp. 14-47.

______.1978. *Grassland Simulation Model.* New York, Heidelberg, Berlin: Springer-Verlag.

______.1979. A spiral approach to ecosystem simulation. In *Systems Analysis of Ecosystems,* ed. G.S. Innis and R.V. O'Neill. Statistical Ecology, Vol. IX, pp. 211-386. Fairland, Maryland: International Cooperative Published House.

Innis, G.S. and O'Neill, R.V., eds. 1979. *Systems Analysis of Ecosystems.* Statistical Ecology, Vol. 9. Fairland, Maryland: International Cooperative Publishing House.

Ito, Y. 1980. *Comparative Ecology.* Cambridge: Cambridge University Press.

Jeffers, J.N.R. 1978. *An Introduction to Systems Analysis: with ecological applications.* London: Edward Arnold.

Jensen, K. and Wirth, N. 1975. *PASCAL: User Manual and Report.* 2nd ed. New York, Heidelberg and Berlin: Springer-Verlag.

Johnson, C.G. 1969. *Migration and Dispersal of Insects by Flight.* London: Methuen.

Jones, D.D. 1977. Catastrophe theory applied to ecological systems. *Simulation* **30**: 1-15.

Jones, D.D. and Walters, C.J. 1976. Catastrophe theory and fisheries regulation. *J. Fish. Res. Bd. Canada* **33**: 2829-33.

Kendall, M.G. and Stuart, R. 1969. *The Advanced Theory of Statistics.* Vol. I, 3rd ed. London: Griffin.

Kenyon, K.W. and Schaffer, V.B. 1954. *A Population Study of the Alaska Fur-seal Herd.* U.S. Fish and Wildl. Service, Special scientific report: Wildlife No. 12, 77pp.

Kettle, D.S. 1951. The spatial distribution of *Culicoides impuctatus* Goet. under woodland and moorland conditions and its flight range under woodland. *Bull. ent. Res.* **42**: 239-91.

Kitching, R.L. 1971a. A simple simulation model of dispersal of animals among units of discrete habitats. *Oecologia* (Berl.) **7**: 95-116.

______.1971b. An ecological study of water-filled tree-holes and their position in the woodland ecosystem. *J. Anim. Ecol.* **40**: 281-302.

______.1972a. The immature stages of *Dasyhelea dufouri* Laboulbene (Diptera: Ceratopogonidae) in water-filled tree-holes. *J. Ent.*(A) 47: 109-14.

______.1972b. Population studies of the immature stages of the tree-hole midge, *Metriocnemus martinii* Thienemann (Diptera: Chironomidae). *J. Anim. Ecol.* **41**: 53-62.

______.1977. Time, resources and population dynamics in insects. *Aust. J. Ecol.* **2**: 31-42.

Kitching, R.L. and Zalucki, M.P. 1982. Component analysis and modelling of the movement process: analysis of simple tracks. *Res. Popul. Ecol.* **24**: 224-38.

Kowal, N.E. 1971. A rationale for modelling dynamic ecological systems. In *Systems Analysis and Simulation in Ecology,* ed. B.C. Patten, Vol. I, pp. 123-97.

Krebs, C.J. 1972. *Ecology: The Experimental Analysis of Distribution and Abundance.* New York: Harper and Row.

Lack, D. 1966. *Population Studies of Birds.* Oxford: Oxford University Press.

Laevestu, T. and Favorite, F. 1977. *Preliminary report on dynamical numerical marine ecosystem model (DYNUMES II) for eastern Bering Sea.* U.S. Natl. mar. Fish Serv., Northwest and Alaska Fisheries Center, Seattle.

Larkin, P.A. 1971. Simulation studies of the Adams River sockeye salmon (*Oncorhynchus narka*). *J. Fish. Res. Bd. Canada.* **28**: 1493-1502.

Larkin, P.A. and Gazey, W. 1981. Applications of ecological simulation models to management of tropical multispecies fisheries. *ICLARM/CSIRO Workshop on the Theory and Management of Tropical Multispecies Stocks*, Cronulla, Australia.

Laws, R.M. 1962. Some effects of whaling on the southern stocks of baleen whales. In *The Exploitation of Natural Animal Populations,* ed. E.D. LeCren and M.W. Holdgate. Oxford: Blackwell Scientific Publications, pp. 242-59.

Leighton, W. 1963. *Ordinary Differential Equations.* Belmont, California: Wadsworth.

Leslie, P.H. 1945. On the use of matrices in certain population mathematics. *Biometrika* **33**: 183-212.

______.1948. Some further notes on the use of matrices in population mathematics. *Biometrika* **35**: 213-45.

Levine, R.D. 1982. Supercomputers. *Scient. American* **246**: 112-25.

Lewontin, R.C. 1969. The meaning of stability. In *Diversity and Stability in Ecological Systems,* ed. G.M. Woodwell and H.H. Smith, pp. 13-24.

Li, J.C.R. 1964. *Statistical Inference II.* Ann Arbor: Edwards Brothers.

Lotka, A.J. 1925. *Elements of Mathematical Biology.* New York: Dover Publications.

Lowe, V.P.W. 1969. Population dynamics of the red deer (*Cervus elaphus* L.) on Rhum. *J. Anim. Ecol.* **38**: 425-57.

Lugo, A.E., Sell, M. and Snedaker, S.C. 1976. Models of terrestrial ecosystems. In *Systems Analysis and simulation in ecology,* ed. B.C. Patten, Vol. IV, pp. 114-46.

Malthus, R.T. 1978. *An Essay on the Principle of Population as it affects the future improvement of Society*. London: J. Johnson.

Margalef, D.R. 1957. La teoria de informacion en ecologia. *Proc. Real Acad. Cien. Artes. Barcelona* **23**: 373-447. Translated by W. Hall in *Gen. Syst.* **3**: 36-71.

Matis, J.H., Patten, B.C. and White, G.C., eds. 1979. *Compartmental Analysis of Ecosystem Models.* Statistical Ecology, Vol. 10. Fairland, Maryland: International Cooperative Publishing House.

May, R.M. 1973. *Stability and Complexity in Model Ecosystems.* Princeton: Princeton University Press.

______.1975. Biological populations obeying difference equation: stable points, stable cycles and chaos. *J. theor. Biol.* **49**: 511-24.

______.1981a. Models for single populations. In *Theoretical Ecology: Principles and Applications,* ed. R.M. May, 2nd ed., pp. 5-29.

______.1981b. Models for two interacting populations. In *Theoretical Ecology: Principles and Applications,* ed. R.M. May, 2nd ed., pp. 78-104.

______.1981c. *Theoretical Ecology: Principles and Applications.* 2nd ed. Oxford: Blackwell.

Maynard Smith, J. 1976. Evolution and the theory of games. *Amer. Scientist* **64**: 41-45.

McCracken, D.D. 1962. *A Guide to ALGOL Programming.* New York: Wiley.

______.1972. *A Guide to FORTRAN IV Programming.* 2nd ed. New York: Wiley.

Meadows, D.H., Meadows, D.L., Randers, J. and Behrens, W.H. III. 1972. *The Limits to Growth.* London: Earth Island Ltd.

Mertz, D.B. 1969. Age-distribution and abundance in populations of flour beetles. I. Experimental studies. *Ecol. Monogr.* **39**: 1-31.

Mesarovic, M. and Pestel, E. 1975. *Mankind at the Turning Point.* London: Hutchinson.

Milsum, J. 1966. *Biological Control Systems Analysis.* New York: McGraw-Hill.

Mitsch, W.J. 1976. Ecosystem modelling of waterhyacinth management in Lake Alice, Florida. *Ecol. Modelling* **2**: 69-89.

Mulholland, R.J. and Sims, C.S. 1976. Control theory and the regulation of ecosystems. In *Systems analysis and simulation in ecology,* ed. B.C. Patten, Vol. IV, pp. 373-89.

Nicholson, A.J. and Bailey, V.A. 1935. The balance of animal populations — Part I. *Proc. zool. Soc. Lond.* **3**: 551-98.

Niven, B.S. 1967. The stochastic simulation of *Tribolium* populations. *Physiol. Zool.* **40**: 67-82.

______.1980. The formal definition of the environment of an animal. *Aust. J. Ecol.* **5**: 37-46.

Odum, E.P. 1953. *Fundamentals of Ecology.* 1st ed. Philadelphia: W.B. Saunders.

Odum, H.T. 1957. Trophic structure and productivity of Silver Springs. *Ecol. Monogr.* **27**: 55-112.

______.1971. *Environment, Power and Society.* New York: Wiley-Interscience.

______.1972. An energy circuit language for ecological and social systems: its physical basis. In *Systems Analysis and Simulation in Ecology,* ed. B.C. Patten, Vol. II, pp. 140-212.

Odum, H.T. and Odum, E.P. 1955. Trophic structure and productivity of a Windward coral reef community on Eniwetok Atoll. *Ecol. Monogr.* **25**: 291-320.

Olson, J.S. 1960. *Health Physics Annual Reports.* Oak Ridge, Tenn.: Oak Ridge National Lab.

O'Neill, R.V. 1979. A review of linear compartment analysis in ecosystem science. In *Compartmental Analysis of Ecosystem Models,* ed. J.H. Matis, B.C. Patten and G.C. White. Statistical Ecology, Vol. IX, pp. 3-28. Fairland, Maryland: International Cooperative Publishing House.

Paris, O.H. 1965. Vagility of p32-labelled isopods in grassland. *Ecol.* **46**: 635-48.

Park, T. and Frank, M.B. 1948. The fecundity and development of the flour beetles *Tribolium confusum* and *Tribolium castaneum,* at three constant temperatures. *Ecology* **29**: 368-74.

Park, T., Mertz, D.B., Grodzinski and Prus, T. 1965. Cannibalistic predation in populations of flour beetles. *Physiol. Zool.* **38**: 298-321.

Patten, B.C. 1969. Ecological systems analysis and fisheries science. *Trans. Am. Fish. Soc.* **98**: 570-81.

______.1971. A primer for ecological modelling and simulation with analog and digital computers. In *Systems Analysis and Simulation in Ecology*, ed. B.C. Patten, Vol. 1, pp. 3-121.

______.Patten, B.C., ed. 1971, 1972, 1975, 1976. *Systems Analysis and Simulation in Ecology.* Volumes I, II, III and IV. New York and London: Academic Press.

Patten, B.C., Bosserman, R.W., Finn, J.T. and Cole, W.G. 1976. Propagation of cause in ecosystems. In *Systems Analysis and Simulation in Ecology,* ed. B.C. Patten, Vol. IV, pp. 458-580.

Paulik, G.T. 1969. Computer simulation models for fisheries research management and teaching. *Trans. Am. Fish. Soc.* **98**: 551-59.

Pearl, R. 1925. *The Biology of Population Growth.* New York: Knopf.

Pennycuick, C.J., Compton, R.M. and Beckingham, L. 1968. A computer model for simulating the growth of a population, or of two interacting populations. *J. theor. Biol.* **18**: 316-29.

Pennycuick, L. 1969. A computer model of the Oxford great tit population. *J. theor. Biol.* **22**: 381-400.

Peterman, R.M. 1977. Graphical evaluation of environmental management options: examples from a forest-insect pest system. *Ecol. Modelling* **3**: 133-48.

Pianka, E.R. 1981. Competition and niche theory. In *Theoretical Ecology,* ed., R.M. May. 2nd ed. Oxford: Blackwell Scientific Publications, pp. 167-96.

Pielou, E.C. 1977. *Mathematical Ecology.* New York: Wiley-Interscience.

______.1981. The usefulness of ecological models: a stock-taking. *Quarterly Review of Biology* **56**: 17-31.

Pollard, J.H. 1966. On the use of the direct matrix product in analysing certain stochastic population models. *Biometrika* **53**: 397-415.

Popper, K.R. 1963. *Conjectures and Refutations.* London and Henley: Routledge and Kegan Paul.

Price, P.W. 1975. *Insect Ecology.* New York: Wiley Interscience.

Ralston, A. 1971. *Introduction to Programming and Computer Science.* Tokyo: McGraw-Hill.

Reichle, D.E., ed. 1981. *Dynamic Properties of Forest Ecosystems.* Cambridge: Cambridge University Press.

Room, P.M. 1979. A prototype "on-line" system for management of cotton pests in the Namoi Valley, New South Wales. *Prot. Ecol.* **1**: 245-64.

Roseberry, J.L. 1979. Bobwhite population responses to exploitation: real and simulated. *J. Wildl. Manag.* **43**: 285-305.

Rosen, R. 1970. *Dynamical Systems Theory in Biology.* New York: Wiley.

Roughgarden, J. 1979. *Theory of Population Genetics and Evolutionary Ecology: An Introduction.* New York: Macmillan.

Ruesink, W.G. 1976. Status of the systems approach to pest management. *Ann. Rev. Ent.* **22**: 27-44.

Russell, C.S., Ed. 1975. *Ecological Modelling in a Resource Management Framework.* Washington: Resources for the Future, Inc.

Sasaba, T. and Kiritani, K. 1975. A systems model and computer simulation of the green rise leafhopper populations in control programmes. *Res. Popul. Ecol.* **16**: 231-44.

Sauer, R.H. 1978. A simulation model for grassland primary producer phenology and biomass dynamics. In *Grassland Simulation Model,* ed. G.S. Innis. New York, Heidelberg, Berlin: Springer-Verlag, pp. 55-87.

Seber, G.A.F. 1973. *The Estimation of Animal Abundance and related Parameters.* London: Griffin.

Shannon, R.E. 1975. *System Simulation: The Art and Science.* Englewood Cliffs, N.J.: Prentice-Hall.

Shannon, C.E. and Weaver, W. 1949. *The Mathematical Theory of Communication.* Urbana: University of Chicago Press.

Shields, P.C. 1968. *Elementary Linear Algebra.* New York: Worth.

Shoemaker, C. 1973a. Optimization of agricultural pest management. I. Biological and mathematical background. *Math. Biosci.* **6**: 143-75.

______.1973b. Optimization of agricultural pest management. II. Formulation of a control model. *Math. Biosci.* **17**: 357-65.

______.1973c. Optimization of agricultural pest management. III. Results and extensions of a model. *Math. Biosci.* **18**: 1-22.

Shugart, H.H. and Noble, I.R. 1981. A computer model of succession and fire response of the high altitude *Eucalyptus* forest of the Brindabella Range, Australian Capital Territory. *Aust. J. Ecol.* **6**: 149-64.

Shugart, H.H. and O'Neill, R.V., eds. 1979. *Systems Ecology.* Strandsberg, Pennysylvania: Dowden, Hutchinson and Ross.

Silliman, R.P. 1969. Analog computer simulation and catch forecasting in commercially fished populations. *Trans. Am. Fish Soc.* **98**: 560-69.

Sinclair, A.R.E. 1973. Regulation and population models for a tropical ruminant. *E. Afr. Wildl. J.* **11**: 307-16.

Skellam, J.G. 1951. Random dispersal in theoretical populations. *Biometrika* **38**: 196-218.

______.1973. The formulation and interpretation of mathematical models of diffusionary processes in population biology. In *The Mathematical Theory of the Dynamics of Biological Populations,* ed. M. Bartlett and R.W. Hiorns. London and New York: Academic Press, pp. 63-85.

Solomon, M.E. 1949. The natural control of animal populations. *J. Anim. Ecol.* **18**: 1-35.

______.1969. *Population Dynamics.* London: Edward Arnold.

Southward, G.M. 1968. A simulation of management strategies in the Pacific halibut fishery. *Int. Pac. Halibut Comm. Report 47.*

Southwood, T.R.E. 1966. *Ecological Methods with particular reference to the study of Insect Populations.* London: Methuen.

______.1977. Habitat, the templet for ecological strategies? *J. Anim. Ecol.* **46**: 337-65.

Spain, J.D. 1982. *"BASIC" Microcomputer Modelling in Biology.* London: Addison-Wesley.

Stark, R.W. 1973. The systems approach to insect pest management — a developing programme in the United States of America: the pine bark beetles. In *Insects: Studies in Population Management,* ed. P.W. Geier, L.R. Clark, D.J. Anderson and H.A. Nix. *Mem. ecol. Soc. Aust.*, No. 1, pp. 265-73.

Steinhorst, R.K., Hunt, H.W., Innis, G.S. and Haydock, K.P. 1978. Sensitivity Analysis of the ELM Model. In *Grassland Simulation Model,* ed. G.S. Innis. Heidelberg, Berlin: Springer-Verlag, pp. 231-55.

Stinner, R.E., Rabb, R.L. and Butler, G.D. 1974. Population dynamics of *Heliothis zea* (Boddie) and *H. virescens* (F) in North Carolina: a simulation model. *Environ. Entomol.* **3**: 163-68.

Tansley, A.G. 1935. The use and abuse of vegetational concepts and terms. *Ecology* **16**: 284-307.

Tassel, D. van 1974. *Program Style, Design, Efficiency, Debugging and Testing*. Englewood Cliffs, N.J.: Prentice Hall.

Taylor, F. 1981. Ecology and evolution of physiological time in insects. *Am. Nat.* **117**: 1-23.

Thompson, W.A. and Vertinsky, I. 1975. Application of Markov chains to analysis of a simulation of birds' foraging. *J. theor. Biol.* **53**: 285-307.

Thompson, W.R. 1924. La theorie mathematique de l'action des parasites entomophages et le facteur de hasard. *Annls. Fac. Sci. Marseilles* **2**: 69-89.

Tillman, M.F. and Stadelman, D. 1976. Development and example—application of a simulation model to the northern anchovy fishery. *Fish. Bull.* **74,** 118–230.

Tomovic, R. 1963. *Sensitivity Analysis of Dynamic Systems.* New York: McGraw-Hill.

Usher, M.B. 1972. Developments in the Leslie Matrix model. In *Mathematical Models in Ecology,* ed. J.N.R. Jeffers, pp. 29–60, Oxford, London, Edinburgh, and Melbourne: Blackwell Scientific Publications.

______. 1976. Extensions to models used in renewable resource management, which incorporate an arbitrary structure. *J. Env. Manag.* **4**: 123–40.

Varley, G.C. and Gradwell, G.R. 1970. Recent advances in insect population dynamics. *Ann. Rev. Ent.* **15**: 1–24.

Vinberg, G.C. and Anisimov, S.O. 1967. Mathematical models of an aquatic system. In *Photosynthesis of Productive Systems.* ed. A.A. Nichiporovich. Jerusalem: Israel Program for Scientific Translation.

Volterra, V. 1926. Variazioni e fluttuazioni del numero d'individui in speci animali conviventi. *Mem. Acad. Lincei* **2**: 31–113. Translation in an Appendix to Chapman's *Animal Ecology,* New York, 1931.

Waide, J. and Webster, J. 1976. Engineering systems analysis: applicability to ecosystems. In *Systems Analysis and Simulation in Ecology,* ed. B.C. Patten, Vol. IV, pp. 330–72.

Walters, C.J. 1969. A generalized computer simulation model for fish population studies. *Trans. Am. Fish. Soc.* **98**: 505–12.

Walters, C.J. and Bandy, J. 1972. Periodic harvest as a method for increasing big game yields. *J. Wildl. Manag.* **36**: 128–34.

Walters, C.J. and Hilborn, R. 1976. Adaptive control of fishing systems, *J. Fish Res. Bol. Canada.* **33**: 145–59.

Walters, C.J., Hilborn, R. and Peterman, R. 1975. Computer simulation of barren-ground caribou dynamics. *Ecol. Modelling* **1**: 303–15.

Walters, C.J. and Peterman, R.M. 1974. A systems approach to the dynamics of spruce budworm in New Brunswick. *Quaest. ent.* **10**: 177–86.

Watt, K.E.F. 1961. Mathematical models for use in insect pest control. *Can. Ent. Suppl.* **19**: 1–62.

______, ed. 1966. *Systems Analysis in Ecology.* New York. Academic Press.

______. 1968. *Ecology and Resource Management.* New York. McGraw-Hill.

Webster, A. 1981/82. *Australian Microcomputer Handbook.* Chatswood, Vic.: 2nd Computer Reference Guide.

Weidhaas, D.E. 1974. Simplified models of population dynamics of mosquitoes related to control technology. *J. econ. Ent.* **67**: 620–24.

Welch, S.M. and Croft, B.A. 1979. *The Design of Biological Monitoring Systems for Pest Management.* Wageningen: PUDOC.

Wiegert, R.G. 1975. Simulation modelling of the algal-fly components of a thermal ecosystem: effects of spatial heterogeneity, time delays and model condensation. In *Systems Analysis and Simulation in Ecology,* ed. B.C. Patten, Vol. III, 1957–82.

Williams, F.M. 1971. Dynamics of microbial populations, In *Systems Analysis and Simulation in Ecology,* ed. B.C. Patten, Vol. I, 198–268.

Williamson, M.H. 1959. Some extensions of the use of matrices in population theories. *Bull. math. Biophysics* **21**: 261–63.

______. 1972. *The Analysis of Biological Populations.* London: Edward Arnold.

Wilson, E.O. and Bossert, W.H. 1971. *A Primer in Population Biology,* Stamford, Conn.: Sinauer.

Woodmansee, R.G. 1978. Critique and analyses of the grassland ecosystem model ELM. In *Grassland Simulation Model,* ed. G.S. Innis. New York, Heidelberg, Berlin: Springer Verlag, 257–81.

Zalucki, M.P. and Kitching, R.L. 1982. Component analysis and modelling of the movement process: the simulation of simple tracks. *Res. Popul. Ecol.* **24:** 239-49.

Zeeman, E.C. 1976. Catastrophe theory. *Sci. Am.* **234**: 65–83.

Index